Clinical Radiotherapy Physics with MATLAB®

A Problem-Solving Approach

Series in Medical Physics and Biomedical Engineering

Series Editors: John G. Webster, E. Russell Ritenour, Slavik Tabakov, and Kwan-Hoong Ng

Clinical Radiotherapy Physics with MATLAB®

A Problem-Solving Approach

Pavel Dvorak

CRC Press
Taylor & Francis Group
Boca Raton London New York

CRC Press is an imprint of the
Taylor & Francis Group, an **informa** business

CRC Press
Taylor & Francis Group
6000 Broken Sound Parkway NW, Suite 300
Boca Raton, FL 33487-2742

First issued in paperback 2020

© 2018 by Taylor & Francis Group, LLC
CRC Press is an imprint of Taylor & Francis Group, an Informa business

No claim to original U.S. Government works

ISBN-13: 978-0-367-57148-1 (pbk)
ISBN-13: 978-1-4987-5499-6 (hbk)

Library of Congress Cataloging-in-Publication Data

Names: Dvorak, Pavel (Medical physicist), author.
Title: Clinical radiotherapy physics with MATLAB : a problem-solving approach
/ Pavel Dvorak.
Other titles: Series in medical physics and biomedical engineering.
Description: Boca Raton, FL : CRC Press, Taylor & Francis Group, [2018] |
Series: Series in medical physics and biomedical engineering | Includes
bibliographical references and index.
Identifiers: LCCN 2018001537| ISBN 9781498754996 (hardback ; alk. paper) |
ISBN 1498754996 (hardback ; alk. paper)
Subjects: LCSH: Medical physics--Data processing. | MATLAB.
Classification: LCC R895 .D86 2018 | DDC 610.1/53--dc23
LC record available at https://lccn.loc.gov/2018001537

Visit the Taylor & Francis Web site at
http://www.taylorandfrancis.com

and the CRC Press Web site at
http://www.crcpress.com

I dedicate this book to my late mother who always encouraged me to do better.

_____

Contents

Foreword

Clinical Radiotherapy Physics with MATLAB®: A Problem-Solving Approach by Pavel Dvorak is the first in the *Series in Medical Physics and Biomedical Engineering* dedicated to software used by clinical medical physicists.

Given the ongoing and accelerating development of medical device technology and user protocols, fulfilling the mission and delivering the full competence profile of the clinical medical physicist has become a daunting task (Caruana et al. 2014; Guibelalde et al. 2014). However, well-written software and programming skills can help in a multitude of ways (Ferris et al. 2005; Lyra et al. 2011; Donini et al. 2014; Nowik et al. 2015). Regrettably, few suitable texts are available, with the result that the acquisition of high-level programming skills by students and clinical scientists is often a hit-and-miss affair. Few books include exemplar scripts illustrating application to the clinical milieu whilst didactic approaches are insufficiently comprehensive or low in communicative power. *Clinical Radiotherapy Physics with MATLAB®: A Problem-Solving Approach* aims to fill this gap with the use of MATLAB® in clinical radiotherapy physics.

In this book, the university tutor will find structured teaching text and real-world case study examples with which to enhance presentations and to set as learning tasks. On the other hand, the student will find a pedagogically appealing and engaging manuscript for individual study whilst the practicing clinical medical physicist will find a learning tool for further development of his or her own skills. Dr Dvorak has utilised a refreshing didactic approach in which he not only provides practical case studies and solutions but also explains the thought processes by which that particular solution was achieved. This approach will help make the acquisition of necessary programming skills more easily accessible to all.

Dr Dvorak is a busy clinical radiotherapy physicist with many years of experience and I sincerely thank him for finding the time within his busy schedule to dedicate to this important educational initiative and to share his radiotherapy and MATLAB expertise with the readers of this book. I am sure it has not been easy and I appreciate it.

Finally, I would like to wish readers of this text many happy programming hours – and to remind them that programming is power!

Carmel J. Caruana, PhD, FIPEM
Professor and Head, Medical Physics Department, University of Malta
Past-Chair, Education and Training Committee, European Federation of Organizations for
Medical Physics
Associate Editor for Education and Training: Physica Medica – European Journal of
Medical Physics
Member Accreditation Committee: International Medical Physics Certification Board

REFERENCES

Caruana CJ, Christofides S, Hartmann GH. (2014). European Federation of Organisations for Medical Physics (EFOMP) policy statement 12.1: Recommendations on medical physics education and training in Europe 2014. *Physica Medica: European Journal of Medical Physics* 30(6), 598.

Donini B, Rivetti S, Lanconelli N, Bertolini M. (2014). Free software for performing physical analysis of systems for digital radiography and mammography. *Medical Physics* May, 41(5), 051903.

Ferris MC, Lim J, Shepard DM. (2005). Optimization tools for radiation treatment planning in MATLAB, in: Brandeau ML, Sainfort F, Pierskalla WP (Eds.). *Operations research and health care. International Series in Operations Research & Management Science*, vol 70. Springer, Boston, MA.

Guibelalde E, Christofides S, Caruana CJ, Evans S, van der Putten W (Eds.). (2014). *European guidelines on the medical physics expert (Radiation Protection Series 174).* Publications Office of the European Union, Luxembourg: European Commission.

Lyra M, Ploussi A, Georgantzoglou A. (2011). MATLAB as a tool in nuclear medicine image processing, in *MATLAB – A ubiquitous tool for the practical engineer*, Clara Ionescu (Ed.). InTech, DOI: 10.5772/19999.

Nowik P, Bujila R, Poludniowski G, Fransson A. (2015). Quality control of CT systems by automated monitoring of key performance indicators: A two-year study. *Journal of Applied Clinical Medical Physics* 16(4), 254–265.

Preface

I wrote this book to share my experience of radiotherapy problem solving with MATLAB® gained during years of educational and clinical practice. During my studies, teaching and clinical service I always sought to understand details of a given principle so that I could apply it in my practical professional life. In my job, I often face non-standard situations and problems requiring local, context-specific solutions. Problems that occur every day simply because of the complexity of the technology and procedures, the need for more personalised care, a large variability of patients and their clinical conditions, hardware and software malfunctions, and, of course, human errors. Such localised problems often require programming for a solution and certainly for eventual automation. Programing platforms, such as MATLAB, make this easier to achieve. I do hope that sharing MATLAB scripts, which I created for purposes of testing and problem solving during the past 10 years, together with a description of algorithms and decision processes, will help you, my younger medical physics colleagues, to acquire and apply the presented principles, approaches and skills in a much quicker and more efficient way than I experienced.

In this book, you will find a direct link between clinical radiotherapy data and MATLAB, programming solutions to problems arising directly from practical clinical issues and commented MATLAB scripts working with actual sample clinical data or your own data matching input requirements. No MATLAB experience is assumed and the intent is to provide sufficient information and guidance so that there is a tangible product at the end of a given chapter so the reader can arrive much closer to a solution than possible with presently available generic MATLAB teaching materials. MATLAB was selected as the computing platform mostly because of its natural and simple syntax and large availability of inbuilt functions; it bypasses the need for extensive programming. Example data and scripts are available in the Web Supplements in the eResources and they follow the content structure of the book. Key algorithms and decision points are mostly presented in the form of structured text and sometimes directly as scripts in MATLAB code. Scripts in the Web Supplements are annotated and reflect the structured text descriptions of the associated algorithms described in the book.

The chosen didactic approach of this book can be described as a *thinking-aloud process*. The author seeks to describe his own thinking process while developing the scripts with the aim of making them accessible to the reader. A problem is stated and questions on precisely how the given problem is defined are asked; these are followed by questions regarding what input and output data are required for a computer-based solution, what is the expected variability of the input, possible solutions with accompanying pros and cons, and so on. In this way, I hope that you, the reader, will be persuaded to *use a more structured thinking process*, sometimes thinking 'outside of the box', even for the simplest problems. Not to 'reinvent the wheel' but to always appreciate the fundamental principles underlying any problem and not to simply succumb to adopting 'common practice' or general guidance. Often, when attempting to find a solution to a given problem in the field of radiation oncology physics, I found study materials lacking sufficient detail to produce an effective functioning application. My experience with many books was that they provided rather *wide* introductions to many topics (albeit referring to further reading from the literature) when *deeper* knowledge was what was required. Therefore, in this book, I tried to provide enough background material so that readers would not need to study recommended literature extensively.

The book begins with an introductory chapter on MATLAB essentials followed by a presentation of examples of data types from radiation oncology practice based on dimensionality and its use in elementary applications. The following two chapters deal with DICOM, starting with examples of radiation oncology–specific data reconstruction in MATLAB (image series, volume structures, treatment plans, dose distributions) and ending with examples of possible data modification to solve particular problems. The remaining chapters are each dedicated to the construction of MATLAB scripts for providing programming solutions for selected practical tasks of radiation oncology physics, such as summing up two treatment plans, merging target volumes in static and dynamic coordinates (target tracking), simple three-dimensional dose calculation, comparing dose distributions using gamma analysis, image-based quality assurance of basic mechanical parameters of a conventional medical linear accelerator and computer modeling of a treatment couch for in vivo dosimetry application. The learning-by-real-examples perspective adopted in this book is an efficient way to achieve real practical progress with MATLAB programming in clinical practice.

The way you use this book will depend on your present level of experience using MATLAB as applied to radiation oncology. If you consider this sufficient, you may go directly to the case study, problem-solving chapters. However, if you only have a general knowledge of MATLAB and have not applied it for solving radiation oncology problems, the initial chapters are a must. In this case, I suggest that you go through the chapters sequentially.

Critical thinking and seeing first principles is the way to a real understanding of Medical Physics in general; understanding is the way to problem solving, and successful problem solving makes the difference between average and better.

MATLAB files, including example scripts and data, are available to download at https://www.crcpress.com/9781498754996 (*the password to download these files is* **Cam5!El**).

Pavel Dvorak, PhD

About the Series

The *Series in Medical Physics and Biomedical Engineering* describes the applications of physical sciences, engineering, and mathematics in medicine and clinical research.

The series seeks (but is not restricted to) publications in the following topics:

- Artificial organs
- Assistive technology
- Bioinformatics
- Bioinstrumentation
- Biomaterials
- Biomechanics
- Biomedical engineering
- Clinical engineering
- Imaging
- Implants
- Medical computing and mathematics
- Medical/surgical devices

- Patient monitoring
- Physiological measurement
- Prosthetics
- Radiation protection, health physics, and dosimetry
- Regulatory issues
- Rehabilitation engineering
- Sports medicine
- Systems physiology
- Telemedicine
- Tissue engineering
- Treatment

The *Series in Medical Physics and Biomedical Engineering* is an international series that meets the need for up-to-date texts in this rapidly developing field. Books in the series range in level from introductory graduate textbooks and practical handbooks to more advanced expositions of current research.

The *Series in Medical Physics and Biomedical Engineering* is the official book series of the International Organization for Medical Physics.

THE INTERNATIONAL ORGANIZATION FOR MEDICAL PHYSICS

The International Organization for Medical Physics (IOMP) represents over 18,000 medical physicists worldwide and has a membership of 80 national and 6 regional organizations, together with a number of corporate members. Individual medical physicists of all national member organisations are also automatically members.

The mission of IOMP is to advance medical physics practice worldwide by disseminating scientific and technical information, fostering the educational and professional development of medical physics and promoting the highest quality medical physics services for patients.

A World Congress on Medical Physics and Biomedical Engineering is held every three years in cooperation with International Federation for Medical and Biological Engineering (IFMBE) and International Union for Physics and Engineering Sciences in Medicine (IUPESM). A regionally based international conference, the International Congress of Medical Physics (ICMP) is held between world congresses. IOMP also sponsors international conferences, workshops and courses.

The IOMP has several programmes to assist medical physicists in developing countries. The joint IOMP Library Programme supports 75 active libraries in 43 developing countries, and the Used Equipment Programme coordinates equipment donations. The Travel Assistance Programme provides a limited number of grants to enable physicists to attend the world congresses.

IOMP co-sponsors the *Journal of Applied Clinical Medical Physics*. The IOMP publishes, twice a year, an electronic bulletin, *Medical Physics World*. IOMP also publishes e-Zine, an electronic news letter about six times a year. IOMP has an agreement with Taylor & Francis for the publication of the *Medical Physics and Biomedical Engineering* series of textbooks. IOMP members receive a discount.

IOMP collaborates with international organizations, such as the World Health Organisations (WHO), the International Atomic Energy Agency (IAEA) and other international professional bodies such as the International Radiation Protection Association (IRPA) and the International Commission on Radiological Protection (ICRP), to promote the development of medical physics and the safe use of radiation and medical devices.

Guidance on education, training and professional development of medical physicists is issued by IOMP, which is collaborating with other professional organizations in development of a professional certification system for medical physicists that can be implemented on a global basis.

The IOMP website (www.iomp.org) contains information on all the activities of the IOMP, policy statements 1 and 2 and the 'IOMP: Review and Way Forward' which outlines all the activities of IOMP and plans for the future.

Acknowledgements

This book would have been much harder to write without the significant support of my close collaborators, friends and family.

Carmel J. Caruana is acknowledged for invaluable help with language, presenting strategy and content review.

A big thank you to Vaclav Spevacek for content review, creating the maths underpinning the points-based registration algorithm and boosting my motivation.

Many thanks to Ivo Chvatil for the front cover design suggestion.

And, of course, a big thank you to my wife for support and endless patience.

Acronyms and Abbreviations

4DCT	Four-Dimensional Computed Tomography Image
ASCII	American Standard Code for Information Interchange
BED	Biologically Effective Dose
CAX	Central Axis (radiation beam)
CBCT	Cone Beam Computed Tomography
CT	Computed Tomography
CTV	Clinical Target Volume
DICOM	Digital Imaging and Communication in Medicine
DRR	Digitally Reconstructed Radiograph
DTA	Distance to Agreement
DVH	Dose-Volume Histogram
EPID	Electronic Portal Imaging Device
EPL	Equivalent Path Length
FDA	U.S. Food and Drug Administration
FFF	Flattening Filter Free
FOV	Field of View
FWHM	Full Width at Half Maximum
GUI	Graphical User Interface
HIS	Hospital Information System
HU	Hounsfield Unit
IGRT	Image-Guided Radio Therapy
IMRT	Intensity Modulated Radiotherapy
ITV	Internal Target Volume
LQ	Linear-Quadratic (model)
MLC	Multileaf Collimator
MU	Monitor Unit
OAR	Organ at Risk/Off-Axis Ratio
OIS	Oncology Information System
PTV	Planning Target Volume
QA	Quality Assurance
QC	Quality Control
ROI	Region of Interest
RS	Radiotherapy Structures

RT	Radiotherapy
R&V	Record & Verify System
SAD	Source Axis Distance
SDD	Source Detector Distance
SIB	Simultaneous Integrated Boost
SSD	Source Surface Distance
TPS	Treatment Planning System
VOI	Volume of Interest

Definition of Terms

The definitions of terms below do not represent the most exact definition possible. Rather they emphasize the key aspects relevant to the context of this book.

4D Computed Tomography

One of the possible approaches to motion management in radiotherapy mostly covers respiratory motion. Using 4DCT extends the standard patient 3D model by another dimension to account for variations of tumor location, tumor shape and surrounding anatomy. Technically, 4DCT is a series of multiple standard 3D CTs acquired at different motion (breathing) phases. One of the possible ways of viewing 4DCT is as a movie loop. In this book, 4DCT is presented as an example of sorting CT images and image series based on DICOM attributes.

Biologically Effective Dose

BED is the radiation quantity used in radiotherapy taking tissue radiosensitivity into account and it is expressed by α, β (LQ model) parameters, dose per fraction, total dose, or overall treatment time, to calculate the biologic effect on a given tissue. Examples of simple dose-to-dose summing two treatment plans (dose distributions) are presented in this book. BED is mentioned in a discussion about the limitation of simplified dose-to-dose plan summations.

Central (beam) Axis

CAX provides a radiation beam geometry reference. Technically, this can be determined using assessment of collimator rotation and geometric center of beam collimation device. Examples of methods determining this crucial parameter for linac QA are demonstrated in this book.

Collimator

Collimator assembly for a conventional linac consists of many components of which the most important are beam collimation devices: 2 pairs (X & Y) of independent secondary jaws generally giving a beam a rectangular shape and MLC. Collimator jaws provide additional shielding for MLC-shaped fields, eliminating interleaf leakage and transmission in the area outside of the intended field. They can be used for beam shaping without the MLC. In dynamic mode, they can modulate beam fluence linearly in the direction of their travel, creating a wedge-like effect on dose distribution in the patient body. In this book, QA tests of collimator rotation axis walkout and scale and jaw position scales are demonstrated as examples of a (semi)-automated QA program based on dedicated image analysis.

Commissioning

Commissioning in radiotherapy is a process of introduction of new equipment in clinical use. It is recommended that this should be a formal and well-structured procedure that includes comprehensive tests of intended application. Commissioning should be well documented. A typical example of commissioning of radiotherapy equipment is a linac and/or TPS. Basically, the treatment machine characteristics must be in agreement with the corresponding beam model in the TPS in order to perform accurate and relevant dose calculation. In this book, some practical aspects of commissioning following acceptance and preceding routine periodic QA are discussed.

Computed Tomography

CT is a three-dimensional imaging method based on mathematical reconstruction of acquisition data using computers. In this book, CT equals X-ray CT, which is based on the difference in photon attenuation by tissues of different electron density. Together with the fact that it is the most available 3D imaging modality, this electron density weighted image nature is the main reason why the CT image series is used for the patient 3D model for radiation treatment planning.

Cone Beam CT

Unlike conventional X-ray CT based on fan beam and detector ring, the CBCT is computed tomography using a cone beam and planar detector. This is the standard setup for contemporary conventional linacs where the X-ray source and planar detector are mounted on linac's gantry perpendicular to treatment beam axis. CBCT is used in 3D image-guided radio therapy (IGRT) by comparison with the original planning CT reference. In this book, CBCT is presented as another example of CT images where processing based on DICOM header information can find a useful application.

Coplanar

Coplanar, 'in one plane', beam geometry consists of beams with their CAX in one transversal plane. For conventional linacs, non-coplanar beams are those with non-zero isocentric rotation of the treatment couch. In general, applying non-coplanar beams provides more degrees of freedom to optimize treatment plan, but it is also associated with higher risk of equipment collision, potentially with the patient.

CT Number

CT image voxel values are generally expressed in CT numbers, which are computed during the tomographic reconstruction process. The main application of CT numbers in this book is CT electron density calibration for 1D inhomogeneity correction in the dose calculation example.

Digital Imaging and Communication in Medicine

DICOM is medical imaging and communication standard assuring that products of different manufacturers can work together based on standardized format of input and output data. Understanding DICOM opens the gates to understanding how things work in medicine. DICOM and its radiotherapy-specific modules (DICOM RT) form the core source of information for most of the applications presented in this book.

Digitally Reconstructed Radiograph

DRR is a planar projection image calculated from a 3D voxel model of a patient created using a CT image series. DRRs are reconstructed for a given source and reconstructed plane position. In conventional photon radiotherapy, this is either the MV (treatment beam) or kV (imaging beam) source and the plane perpendicular to the beam axis through a treatment machine isocenter. A given DRR is compared with a corresponding real X-ray image acquired using MV or kV planar detector on a treatment machine, which is then scaled to the same plane position as the calculated reference DRR to verify patient (or target) position.

Distance to Agreement

DTA is a concept used in quantitative comparison of dose distributions, which is important in radiotherapy quality assurance. In this book, the DTA is discussed as a component of the gamma index, the contemporary standard for comparing two dose distributions.

Dose-Volume Histogram

For any given structure, a histogram of absorbed dose on the x-axis and the percentage volume of the structure receiving that particular *or higher* absorbed dose. DVH is a widely accepted quality metric for comparing dose distributions in the patient body in radiotherapy. Risk of radiation-induced toxicity of critical organ is often expressed in terms of DVH statistics. Processing, reconstructing and calculating DVHs are topics included in several chapters of this book.

Electronic Portal Imaging Device

EPID is a planar detector of megavoltage photon beams of conventional medical linear accelerators. Developed originally for early stages of IGRT to verify patient position before treatment, nowadays, it is used more in radiotherapy quality assurance, such as three-dimensional transit dosimetry in vivo. In this book, the application of EPID to measure basic mechanical parameters of a conventional linac is demonstrated in examples.

Equivalent Path Length

Sometimes, also referred as 'water equivalent depth' or 'effective depth' is the concept used in dose calculation for radiation treatment planning. It is the distance in tissue weighted by the relative electron density of that tissue to water. EPL is the simplest 1D inhomogeneity correction using CT images weighted by electron density to account for other than water-like tissue. The EPL concept is applied in a demonstration of 3D dose calculation algorithm presented in this book.

European Council

The European Council, together with the European Parliament issue European regulations. These regulations are called (EC) directives. One of them deals with medical devices. For the purposes of this book, their approach to software and particularly customized software for applications in radiotherapy is the most relevant.

Field of View

This book refers to the *scanned* FOV, the volume from which the acquisition data are acquired. This volume may be the same or larger than the volume from which the image is eventually reconstructed from the acquisition data (known as the reconstruction FOV or displayed FOV). Due to the fixed size of reconstruction matrices, the reconstructed FOV determines the pixels size. Necessity to account for variable pixel size due to nonconstant FOV is an important factor used in example applications in this book.

Flattening Filter Free

FFF photon beams are available in some modern medical linear accelerators. With the introduction of differential dose calculation algorithms, flat beams became unnecessary. The major feature of FFF beams is the relatively higher beam fluence and dose rate with a potential to reduce treatment times, which is an important factor, especially in hypofractionated (stereotactic) radiotherapy. In this book, the FFF beams are used in examples of dose profiles and dose calculation.

Fluence

Unlike intensity, the fluence is a clearly defined physical quantity specifying 'amount of ionizing radiation' by ICRU. Within this book, both terms 'fluence-modulated' and 'intensity-modulated' are used. The first is used for topics where the actual physics nature of the radiation beam requires a description; the second is used solely in association with the less precisely defined vendor established term intensity-modulated radiotherapy (IMRT).

Full Width at Half Maximum

FWHM is a common method to define radiation field size and borders. The most related topic discussed in this book are methods of determining field borders in the chapter dealing with automation of QA procedures for conventional linac.

Gamma

The gamma index is a metric used for quantitative comparison of two dose distributions in radiotherapy physics. It consists of two components: difference in dose and distance to agreement. A practical demonstration of implementing the gamma using MATLAB is the topic of an entire chapter of this book.

Hounsfield Unit

A special case of CT numbers assigning HU to zero for distilled water under reference temperature and pressure and –1000 for air.

Image Frame

Dynamic images consist of a sequence of static frames. Frames can be integrated or averaged, frame rate is one of important parameters of each imaging device describing its time resolution. In this book, frames are referred to in chapters dealing with reconstructing CT images and 3D dose distributions from DICOM data.

Image-Guided Radio Therapy

All contemporary radiotherapy should fall in the category of IGRT, that is, image-based verification of patient position before treatment delivery should be implemented at least in a simplified form of checking patient setup without imaging (offline correction protocol). There are many IGRT technologies and also many approaches available. Verification 3D imaging using CBCT with online soft-tissue based correction of patient position can be considered current IGRT standard.

Image Registration/Fusion

Using secondary images for radiation treatment planning is becoming more common, mostly to guide target definition in situations where single planning CT series is insufficient. Modern TPS allow importing multiple secondary image series of various modalities (CT, MR, PET/CT, DSA). In order to use these images, they must be registered/fused with the primary (CT) image. Registration objective depends on the clinical purpose for which the fused images are required. Target definition is a different purpose than, for example, motion assessment. One of the topics in this book, treatment plan sums, is based on manual rigid registration of two CT images.

Intensity Modulated Radiotherapy

Radiotherapy using beams with modulated fluence. The simplest examples of beam fluence modulators commonly used in contemporary non-IMRT are electronic wedges and field-in-field (combination of multiple static uniform-fluence fields of different weight delivered from the same direction). Radiation beams falling into the IMRT category are modulated in 2D. Fluence modulation in conventional linac is ensured by the MLC. An example modification of the radiation treatment plan to change beam fluence for testing purposes is presented in this book.

Internal Target Volume

Based on the ICRU definition, the ITV is the Clinical Target Volume expanded by its expected changes in position due to internal motion. Merging volumes to create ITV in both static and dynamic coordinate systems is one topic demonstrated in this book.

In Vivo Dosimetry

Independent dosimetry on patients during dose delivery to verify either machine performance or patient position or both. A treatment couch modeling example demonstrated in this book was originally created for application in dosimetry in vivo using point detectors.

Inverse Treatment Planning

When few treatment machine parameters were available to optimize dose distributions, it was possible to produce plans manually by an iterative trial-and-error approach. This is a traditional – *forward* – way of radiation treatment planning. With the introduction of IMRT with orders of magnitude larger numbers of plan parameters, the forward planning approach is no longer possible and computerized optimization takes over through the *inverse* planning process. The inverse aspect consists of defining the parameters of desired dose distribution and the algorithm provides sets of treatment machine parameters – treatment plan – which would be optimal with regard to the specified goals.

Isocenter

Conventional photon linacs belong to category of isocentric treatment machines. Isocenter is the point in space where all major axes of the treatment machine components intersect. This includes gantry and collimator rotation and treatment couch rotation. When the target is placed in the isocenter, this setup allows approaching it from a large number of beam directions. The number of beams and their direction is a significant factor for dose distribution conformity. Testing the isocenter location and stability is demonstrated in this book.

Isotropic

Equivalent in all directions. Demonstration of nonisotropic expansions and reductions of volumes of interest (structures) to create defined margins are presented in one of chapters of this book.

Linac

Conventional medical linear accelerator of electrons producing megavoltage photon beams. Standard accessories include MLC for beam shaping and fluence modulation and kV imaging system for IGRT.

Matrix

In this book, the term matrix is mostly equivalent to 2D- or 3D-array, representing numerical data organized in plane or volume, respectively. Matrices defining a given mathematical transform are used as well.

Monte Carlo

Explicit particle transport modeling approach applied at various levels in modern photon or particle dose calculation algorithms. Calculating absorbed dose using explicit simulations of particle interactions is demanding in resources but provides, in principle, the highest accuracy. Within this book, Monte Carlo is referred to as the contemporary gold standard in dose calculation algorithms used in radiation treatment planning.

Monitor Unit

MUs quantify 'amount of radiation' produced by a linac. It is based on the signal from transit ionization chambers placed in the beam path. Determination of the MUs required to deliver a desired radiation dose in a given point within a patient body is the basic task of radiation treatment planning.

MultiLeaf Collimator

Radiation beam fine shaping device. Consisting of tens of pairs of opposite leaves to shape a given beam. In dynamic IMRT, a traveling leaf pair during beam on creates a nonuniform fluence profile based on the difference in speed between the two leaves. In this book, MLC is used to define beam CAX for automated quality assurance of conventional linac mechanical parameters.

Off-Axis Ratio

Also referred to as Off-Center Ratio is the ratio of dose-off beam CAX and dose-on beam CAX at a given depth for a given radiation beam in reference material (water). Using the OAR (do not confuse with Organ-At-Risk) is the simplest approach to account for beam fluence nonuniformity in dose calculation. The OARs are used in the simple 3D dose calculation example presented in this book.

Oncology Information System

Radiotherapy requires management of large amounts of patient data including, for example, image series, verification images, treatment plans, delivered dose records, appointment scheduling, reporting, follow-up information, and so on. There are commercial software products managing all of these and much more. When oncology is a part of a hospital with a general Hospital Information System, it is desirable that the OIS is connected to the HIS.

Optimization

Optimization in radiotherapy is meant mostly (but not exclusively) in reference to IMRT and inverse planning. Inverse planning relies on the mathematical process of optimization, that is, searching the optimal multiple parameter value combinations, which would lead to the desired dose prescription to the target volumes and organs at risk.

Organ at Risk

Organ at Risk (also referred to as *critical organ* or *critical structure*) is a human organ exposed to risk of radiation damage in radiotherapy. OARs are usually organs within or close to the treated area.

Patient 3D Model

Modern radiotherapy is three-dimensional or even four-dimensional when patient regular motion is considered during treatment planning and/or treatment delivery. Three-dimensional treatment planning requires 3D model of a patient to simulate treatment beams and their parameters and how they translate as doses in the patient's body. Hence, a patient 3D model is essential. X-ray CT of a patient in the treatment position, including all immobilization devices, is the standard imaging modality to provide such a patient 3D model. Planning the CT image series/patient 3D model and related aspects is the key attribute of many examples in this book.

Phantom Physical

An application-specific device is usually made from human tissue–equivalent material to simulate radiation measurements on patients under specific conditions. The most common phantom material is water as a substance with radiation properties very close to human (soft) tissue. Within this book, dedicated phantoms are mentioned as tools required to provide specific measured data for analysis.

Phantom Virtual

Virtual phantoms replace real physical phantoms where possible and where this is a more efficient way to achieve a given goal. Typically, these are artificially generated image series to study a particular effect of a specific software. A possible way of creating a virtual phantom is demonstrated in one of the chapters in this book.

Planning Target Volume

The planning target volume is the major dose optimization objective during radiation treatment planning. An optimum treatment plan results in a dose distribution that covers the PTV by a desired (treatment) dose and minimizes dose to adjacent tissue, especially critical organs (OARs). PTV is used many times within this book.

The example of DICOM RT data given in this book contains the PTV in the structure set; PTV DVHs are presented while PTV is also meant as a target volume for calculating dose distribution quality metrics – conformity index – the topic of one chapter in this book.

Quality Assurance/Control

Learning by example to create scripts in MATLAB processing real clinical data can be efficiently used in validation of, for example, third-party software or designing routine periodic tests of major mechanical parameters of a conventional linac.

Radiotherapy Structures

Radiotherapy structure is the official term used in DICOM RT standard for volumes of interest defined on a given 3D image series, mostly for the purposes of radiation treatment planning. From a technical perspective, a structure is a group of voxels of the patient 3D model that have something in common: they belong to one critical organ or its part, one target volume, a part of the target volume, and so on. RT structures are a crucial object in the whole book. The topics include reconstructions, modifications, volume expansions and reductions, DVH calculations, and so on.

Record & Verify System

The R&V system is a term used commonly for software that ensures the accurate transfer of the radiation treatment plan parameters to a treatment machine. In the past, the output of the radiation treatment planning process was a printed set of treatment machine parameters to deliver the desired dose, which had to be input into the machine manually. Due to orders of magnitude larger number of parameters used today, this is impossible and dedicated software took over, resulting in a much more efficient and safer data transfer. Modern R&V software does much more, including recording the delivered dose.

Region of Interest

ROI is a crucial term used in imaging and also in this book. It defines the subset of pixels (or voxels) in an image specified by some common characteristics, for example, pixels (voxels) defined as part of a given organ shown in the image. In this book, ROIs typically relate to two-dimensional images, whereas VOIs relate to three-dimensional images.

Simultaneous Integrated Boost

SIB is a term for a technique in modern radiotherapy that delivers various doses to multiple target volumes during a single treatment fraction. This approach saves on the total number of fractions a patient has to undergo and also allows more customized treatment. SIB is possible on modern treatment machines capable of delivering highly conformal dose distributions.

Syntax

The set of rules of writing. In computing (writing scripts or commands to execute them) there is zero tolerance on typographic errors so whenever a computer code or its part is presented, this has to be 100% correct per rules specific to the programming language used.

Tongue & Groove

MLC leaves have a tongue-and-groove design to reduce interleaf leakage. This causes slight asymmetry in beam definition in the direction perpendicular to the direction of travel of the leaves. In this book, this can have a negligible effect on algorithms defining field borders and beam CAX.

Treatment Planning System

In a traditional way, a TPS is a computer with a specialist software to perform *treatment planning*. Input data is a patient CT model, radiotherapy structures contoured by a radiation oncologist, prescribed dose, and a given treatment machine model. The output is a *treatment plan*: a set of parameters of the given treatment machine to deliver the desired (optimized) dose distribution to the patient. The core of each TPS is a dose calculation engine for the given radiation modality. A modern TPS includes more features, such as image registration, biological planning algorithms or cloud computing-based, dose calculation and optimization.

Vector

In this book, the term vector is mostly equivalent to 1D-array, representing numerical data organized in line. Vectors defining a given position or direction are used as well.

Volume of Interest

VOIs are crucial objects in radiotherapy and also in this book. Also referred to as radiotherapy structures, they are defined as specific objects in DICOM RT standard and they provide essential geometric information about which parts of the patient 3D model are targets (requiring treatment dose) and which parts are organs at risk (dose as low as possible). They allow calculating DVHs as contemporary standard metrics to quantify dose effects.

U.S. Food and Drug Administration

The FDA approves medical devices to be used in healthcare in the United States. For purposes of this book, the FDA's approach to software and particularly customized software for applications in radiotherapy is the most relevant.

CHAPTER 1

MATLAB Essentials and Principles of Simple Programming

LEARNING OUTCOMES

LO 1.1   Explain the essentials to start working with MATLAB
LO 1.2   Explain the concept and essential principles of scripting in MATLAB
LO 1.3   Explain the essential principles of algorithmisation
LO 1.4   Perform basic input/output operations including data display, saving, export, and import
LO 1.5   Perform essential file and folder management operations in MATLAB
LO 1.6   Explain the importance of problem specification
LO 1.7   Explain the importance of program testing and documentation

ALL YOU NEED TO START

Why MATLAB? MATLAB (The Mathworks Inc.) is an established scientific computing platform that has an easily understandable concept, offers powerful HELP and has a wide community of users. The author has decided to use the MATLAB platform because of its simple and natural syntax and for the wide availability of inbuilt functions that can be used efficiently. Optimized inbuilt functions reduce the need for programming at a more basic level, thus saving time and effort. It is inefficient to write scripts for well-established and clearly defined operations (eg, matrix operations such as multiplications, transpositions, or rotations). MATLAB is available as a basic platform with the possible addition of optional toolboxes. The only toolbox used in this book is the Image Processing Toolbox.

MATLAB Interface

Like many common computer applications today, MATLAB opens after the appropriate icon or start menu item is selected and executed. Normally, MATLAB (version R2013a) opens with the default layout shown as a scheme drawing in Figure 1.1.

The layout can be customized intuitively and easily using the *Layout* icon in the toolstrip. Most of the layout elements are natural and do not require any explanation for today's computer user. The most important MATLAB specific features are:

- The PLOTS and APPS tabs contain interactive applications to help guide the user through some specific activities such as plotting a selected variable or curve fitting

- However, working with MATLAB means giving commands in the Command Window. These commands can call MATLAB inbuilt functions, or, they can call scripts created by the user using the MATLAB (or any other text) editor following the MATLAB syntax

- Most of the commands create variables. All current variables including their basic properties are displayed in the Workspace Window

- Until deleted, commands entered in the Command Window are recorded in the Command History and can be navigated by using the dedicated window in the layout (alternatively, the UP and DOWN arrow key with or without specifying commands to search within by indicating first one or few letters, can be used for quick access to Command History from the Command Window directly)

- Folders and navigation helps in organizing an efficient workflow

- Any current variable can be deleted by the *clear* command in the Command Window

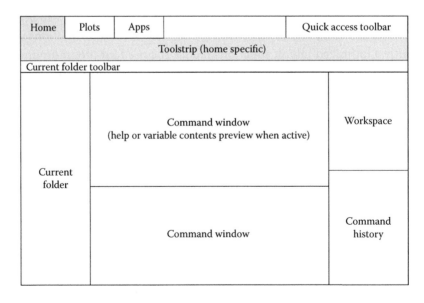

FIGURE 1.1   MATLAB (version R2013a) default layout scheme.

For example:

```
>> clear variable1 variable2
```

Clears two variables: *variable1* and *variable2*.

- *Clear* without specifying variable(s) clears all current variables, ie, whole workspace Alternatively, the *Clear Workspace* icon in the toolstrip can be used

- Any variable can be renamed directly in the Worskpace Window

- To see details about all current variables, use the *whos* command in the Command Window

To see details of specific variables use:

```
>> whos variable1 variable2
```

- To clear the Command Window use the *Clear Commands* icon in the toolstrip

- Scripts (ie, MATLAB programs) are sequences of commands using MATLAB in-built functions. To start scripting (writing your own scripts, ie, MATLAB programs), access the MATLAB Editor by using the *New Script* icon in the toolstrip

- Help and documentation for any MATLAB inbuilt function can be displayed by a respective command in the Command Window

>> help functionName   % display help in the Command Window

>> doc functionName   % display documentation in (newly opened) HELP Window

Alternatively, the HELP Window can be opened directly using the icon in the toolstrip. This is necessary when a given function name is unknown or full text search is required.

Presenting Strategy for This Book

Since this book contains both the text describing the mathematical, physics and clinical aspects of a given subject and associated MATLAB commands and scripts, it is both challenging and important to choose an optimum strategy for presenting the information. Single commands and short command sequences are presented directly in the text. To make reading easier, longer command sequences or functions are presented in tables with step-by-step explanation. Every command in the MATLAB Command Window is preceded by the >> symbol. Where relevant, the MATLAB answer to a given command is shown here, too. Rather than complete information for a given topic, an example with relevance for future application is preferred, since it is not useful to the reader to reproduce MATLAB's high quality and easily accessible Help. The Web Supplement is an integral part of this book. It contains sample data that the text refers to, as well as dedicated functions and scripts to deal with selected problems of radiotherapy physics. Scripts often cannot be presented in the text owing to their length and will therefore be found in the Web Supplement, so that the text may focus on describing the algorithm and the relevant clinical and physics aspects used to make the particular decisions for solving a given problem or its part.

In addition to ease in reading, an uncomplicated approach and solution is also important (as opposed to brevity). Some of the algorithms and solutions presented here may have more computationally efficient alternatives, however, the author believes that the reader can follow the algorithms and information presented here much more easily with a minimum experience with programming and practical application of higher mathematics.

First Steps: Using MATLAB as a Pocket Calculator

The easiest way to start working with MATLAB is to try basic numerical operations as with a pocket calculator. Let's focus on the Command Window and learn the essentials using the commands shown in Table 1.1. The most important learning point here is to understand the concept of variables. Performing an operation without assigning the result to a new, user defined variable assigns this result to the MATLAB default variable *ans*. Only the current result of the last operation is stored in the *ans* variable.

TABLE 1.1   First steps in MATLAB: Basic numeric operations

Command Window	Comments

>>   2+3
        ans =
                    5
	%* simple sum of two numbers (enter). Result assigned automatically to variable *ans* (answer) and printed on-screen as no semicolon at the end of line

>>   ans
        ans =
                    5
	% print content of variable *ans*

>>   a=2
        a =
                    2
	% define variable *a* by assigning the value of 2. Variable content is printed on-screen (no semicolon at the end of line)

>> a=2;	% re-define variable *a* and assign it the value of 2 without printing variable contents on-screen (using semicolon at the end of line)
>> b=3;	% define variable *b* and assign it the value of 3 without printing variable contents on-screen

>>   c=a+b
        c =
                    5
	% calculate sum of variables *a* and *b* and assign the result to a new variable *c* with printing on-screen

>>   pi
        ans =
                3.1416
	% print MATLAB intrinsic constant *pi*. Automatic variable *ans* now changes its content and prints it

>>   sin(45/180*pi)
        ans =
                0.7071
	% calculate sine 45deg (must be converted in radians). Automatic variable *ans* now changes its content and prints it

>>   c=sqrt(a^2+b^2)
        c =
                3.6056
	% define variable *c* as the square root of the sum of squares of variables *a* and *b* with printing on-screen

*Text behind '%' sign is recognized by MATLAB as comments only, ie, MATLAB does not try to execute anything in the line beyond the sign.

First Steps: Vector Algebra and Composite Commands

Since this book is about radiotherapy physics, it is essential to work with vectors and matrices. Table 1.2 demonstrates the essentials of vector definition and basic operations in MATLAB. The most important learning objective is using the selection of vector elements based on logical input and the *find* command and understanding that it returns indices of elements meeting the given search condition.

TABLE 1.2   First steps in MATLAB: Basic vector algebra

Command Window	Comments
>> a1=[-2,-1,0,1,2,3] a1 = -2 -1 0 1 2 3	% define vector *a1* of integers from –2 to 3 by explicit entering values separately
>> a2=[-2:1:3] a2 = -2 -1 0 1 2 3	% define vector *a2* of integers from –2 to 3 as a vector of increasing numbers starting with –2 and ending with 3 with increments of 1
>> a3=[-2:3] a3 = -2 -1 0 1 2 3	% define vector *a3* of integers from –2 to 3 as a vector of increasing numbers starting with –2 and ending with 3 with default increment of 1
>> a4= -2:3 a4 = -2 -1 0 1 2 3	% define vector *a4* of integers from –2 to 3 as a vector of increasing numbers starting with –2 and ending with 3 with default increment of 1
>> a1_4=a1(4) a1_4 = 1	% define new variable *a1_4* equal to value of the 4th element of vector *a1*
>> abs(a1)>1 ans = 1 0 0 0 1 1	% returns logical response to numerical condition: 1 for elements of vector *a1* with absolute value above 1 and 0 for remaining elements
>> a1(abs(a1)>1) ans = -2 2 3	% performs selection of vector *a1* elements based on logical response to numerical condition
>> a1_AbsOver1=find(abs(a1)>1) a1_AbsOver1 = 1 5 6	% define new variable containing vector indices of all elements of vector *a1* with absolute value >1
>> b=a1+a2 b = -4 -2 0 2 4 6	% define new variable *b* as vector sum of vectors *a1* and *a2*
>> c=2*a1-a2 c = -2 -1 0 1 2 3	% define new variable c as vector sum of vectors 2× *a1* and –*a2*
>> a1_2ndTo4th=a1(2:4) a1_2ndTo4th= -1 0 1	% define new variable as selection of elements 2 to 4 of vector *a1*

>> a1_2ndTo4th'
ans =
                     -1
                      0
                      1
	% transpose the last defined vector: from original dimension 1 × 3 to new 3 × 1 (rows × columns)

Note that some basic vector operations demonstrated in Table 1.2 include composite commands where a function output is used as another function input parameter without creating an additional variable. For example:

```
>> a1 (abs(a1>0))
```

is a composite command where a logical output based on a numeric condition is determined first, and is then followed by selecting the relevant vector elements in the next. Composite commands are used efficiently whenever the outcome of an internal function is not explicitly needed for further processing.

Essential matrix definitions (2D) and basic operations are demonstrated in Table 1.3. A major learning point is the importance of dimensionality. Element-by-element matrix operations require matrices of the same dimensions. Matrix multiplication requires appropriate dimensions as stated by operation definition.

TABLE 1.3 First steps in MATLAB: Basic matrix operations

Command Window	Comments
`>> A=ones(2,3)` `A =` `1 1 1` `1 1 1`	% define variable *A*, matrix of ones 2 (rows) by 3 (columns) using MATLAB intrinsic *ones* function
`>> B=[cos(45/180*pi),-sin(45/` `180*pi),0;sin(45/180*pi),` `cos(45/180*pi),0;0,0,1]` `B =` `0.7071 -0.7071 0` `0.7071 0.7071 0` `0 0 1.0000`	% define new variable *B*, matrix 3 × 3, by explicit entering element values. Comma ',' separates elements in one row, semicolon ';' separates rows. Note this is a 2D transformation matrix describing rotation by 45 deg.
`>> B(1:2,:)-A` `ans =` `-0.2929 -1.7071 -1.0000` `-0.2929 -0.2929 -1.0000`	% perform numerical operation using matrices *A* and selection of *B* (to match the dimensions of *A* to enable element-by-element subtraction)
`>> 2*A.*B(2:3,:)` `ans =` `1.4142 1.4142 0` `0 0 2.0000`	% perform element-by-element multiplication between 2× *A* and same size selection of *B*. The result is in the default variable *ans*. Note the operator '.*' is not matrix multiplication but element-by-element multiplication
`>> ans(2,3)` `ans =` `2`	% test of matrix element addressing using current default *ans* variable
`>> B*[1,1,1]'` `ans =` `0.0000` `1.4142` `1.0000`	% perform matrix multiplication between matrix *B* and transposed vector of appropriate dimension to enable matrix multiplication operation ('*'). Note this operation performs rotation by 45 deg of vector [1,1] (first 2 elements)
`>> B*[1,-1,1]'` `ans =` `1.4142` `-0.0000` `1.0000`	% same operation as above for different vector [1,–1]

First Steps: Data Plots and Figures

Table 1.4 demonstrates essential plotting and figure(s) handling commands using the *sin(x)* function. The command sequence results in the graph as shown in Figure 1.2.

First Steps: Program Development (Scripting)

It is time now to try our first program/script/code for calculating a cylinder volume. As we know there are two parameters for determining a cylinder: the radius of the base (*r*) as well as the height (*h*), and cylinder volume (*V*), calculates then as:

$$V = \pi \times r^2 \times h$$

Naturally, if we know the values of the input parameters (*r* = 2, *h* = 5), in such a simple case, we can just perform one command in MATLAB Command Window and get the result:

TABLE 1.4   Basic data plots

Command Window	Comments
>> x = -2*pi:.1:2*pi;	% define *x* values vector for example graph: range from –2 to 2 with 0.1 spacing
>> y = sin(x);	% define *y(x)* values (vector) for example graph
>> plot(y)	% create new (1st) figure object and plots example graph – *y* values only. *x*-values default as the *y* vector element indices (default solid blue line)
>> figure	% create new empty figure object
>> plot(x, y)	% plots *y* vs. *x* data graph
>> close (figure(1))	% close selected figure object – figure (1)
>> figure(2)	% selects figure (2) as current for next figure operations
>> hold on	% hold data displayed in current figure (figure (2)) to display additional data in the same figure
>> plot(x(1:2:end), y(1:2:end), 'or')	% plot every second [*x,y(x)*] data point as red circles in current figure (2)
>> axis([-pi pi -1.2 1.2])	% define *x* and *y* axis range for current plot
>> hold off	% release data hold for current figure (2)
>> title ('This is example figure title')	% add title …
>> xlabel ('Example x-axis label')	% add *x*-axis (horizontal) label
>> ylabel ('Example y-axis label')	% add *y*-axis (vertical) label
>> legend ('full data', 'every 2nd data')	% add displayed data legend
>> saveas (1, 'exampleFigureA', 'fig')	% save current figure (2) as MATLAB *exampleFigureA.fig* file in current folder
>> saveas (1, 'exampleFigureB', 'jpg')	% save current figure (2) as MATLAB *exampleFigureB.jpg* file in current folder
>> close all	% close all opened figure objects
>> open ('exampleFigure1.fig')	% open MATLAB *.fig* file from current path

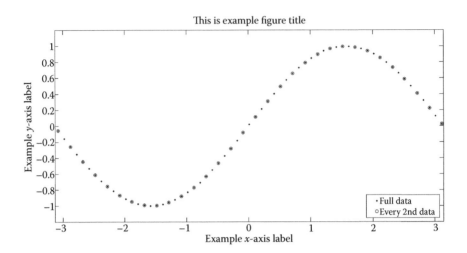

FIGURE 1.2   Plot created by the command sequence in Table 1.4.

```
>> V=pi*2^2*5
V =
   62.8319
```

This approach can hardly be considered programming. However, if we create a new script by using the *New script* icon on the MATLAB toolstrip, MATLAB Editor opens and we can write and save a new script – program.

Type the following text in a new script template.

```
% this is my first MATLAB script...
% just a sequence of commands to run from MATLAB command line
% calculate cylinder volume for fixed r=2 and h=5...
r=2; h=5;
V=pi*r^2*h
```

… and save the script as *cylVol_fixed.m* (in current working folder). Return in MATLAB Command Window and run the new script:

```
>> cylVol_fixed
V =
   62.8319
```

So, the script called *cylVol_fixed* is your first MATLAB program calculating volume of cylinder with fixed $r = 2$ and $h = 5$.

It is obvious that a script like this with fixed parameters has near zero practical value since what it calculates can be done only once. If we want to calculate the cylinder volume for different values of input values r, h then the first option we have is to change the r and h values in the script itself, re-save, and then re-run. Again, this approach does not have a big practical value for simple cases like this, but imagine that your problem is more complex, say with 10 input parameters with a sequence of 50 (rather than 1) commands and calculations. Then, even manual adjusting the values of these 10 parameters – ideally in the beginning of the script to be found easily – re-saving the script and running it to obtain the result, can be considered already as a useful program compared with retyping the whole sequence of commands in MATLAB Command Window (forget the universal CTRL+C -> CTRL+V magic for a moment).

The next step moving us even further towards a 'proper program' using a cylinder volume as a simple example is to replace the manual adjustment of input values within the script by introducing requests to input the desired values directly in the MATLAB Command Window as part of the script. The easiest way of doing this for such a simple example is using the *input* command. Return to MATLAB (script) Editor (or use the *New script* again) and write a modified script:

```
% this is my second MATLAB script...
% sequence of commands including request to enter input values in
MATLAB command line
% calculate cylinder volume for any values entered when prompted
during the script execution...
r = input ('Please, enter the cylinder radius... ');
h = input ('Please, enter the cylinder height... ');
V = pi*r^2*h
```

... and save as *cylVol.m*. Then return to MATLAB Command Window and run the script. Enter desired parameter values when prompted:

```
>> cylVol
Please, enter the cylinder radius... 2
Please, enter the cylinder height... 5
V =
   62.8319
```

Congratulations! You have created a script calculating a cylinder volume for any radius and height specified during the script execution.

Now imagine you would like to use the cylinder volume calculations as part of some bigger script or program. Let's say our task is to calculate the total volume of all cylinder shaped swimming pools in a country and what we have is a list of cylinder radii and associated heights for a thousand pools. Unless paid with a hefty hourly rate, nobody would like to run your *cylVol* script a thousand times, manually entering each pool's parameters, recording the result for each item and summing it up at the end. What we would like to have is parameter data imported in MATLAB as the input variable – perhaps a 2D array [r, h] with 2 columns and 1000 rows – and then a MATLAB *function* calculating one value (cylinder volume) for two values on its input (r, h). Result of each calculation can be saved in another variable, say a 1D array [V] (1 column and 1000 rows) and sum calculated at the end. So, what we need is the input data [r, h], and a function calculating volume to calculate the elements of the intermediate result [V]. Let's create the function – as a new script in MATLAB Editor – first:

```
% this is my third MATLAB script...my first function
% calculate cylinder volume for two parameters specified when
% calling the function

function V = cylVolFunction (r, h)
V = pi*r^2*h
```

Now, having the function ready, let's generate the input data for the swimming pools example by using a random numbers generator. The MATLAB inbuilt function *random* generates vectors of random numbers for specified distributions and parameters. Let's go to the MATLAB Command Window and generate a vector of 1000 rows (1 column) of random numbers from a normal (ie, Gaussian) distribution with mean *2.0* and *0.5* standard deviation; this will be our list of all cylindrical swimming pools radii from the country:

```
>> r = random ('Normal', 2.0, 0.5, 1000, 1);   % produces 1000x1
matrix variable r with random normally distributed values in the
first column
```

Let's do the same for swimming pool heights however, this time with mean *1.5* and *0.5* standard deviation:

```
>> h = random ('Normal', 1.5, 0.5, 1000, 1);
```

Now, let's create a new variable *rh* combining both arrays in one matrix 1000 × 2 (first column for radii, the second for heights) to have just one input variable:

```
>> rh = [r, h];
```

Original *r* and *h* are no longer needed so ... delete them:

```
>> clear r h
```

Now, without generating a new script, let's calculate volumes for all swimming pools whose parameters we have. Let's use the *for* loop command for the first time as well.

```
>> for i = 1:size(rh, 1)
        V(i) = cylVolFunction (rh (i, 1), rh (i, 2));
   end
```

... addressing the *rh* input one-by-one using the index *i* up to the *rh* size in the first dimension (rows) calculate corresponding cylinder volume *V(i)* using the function for given radius *r = rh (i, 1)* and height *h = rh (i, 2)*.

Calculate the resulting total volume by summing up all the *V* vector elements.

```
>> totalVol = sum(V)
totalVol =
   2.0797e+04
```

Note that the *V* variable changes its value in every loop during the script execution. Although allowed, this is a computationally inefficient process in MATLAB and pre-allocating a variable is recommended. Pre-allocating is particularly easy when the variable size is known such as in this example (1000 × 1):

```
>> V = zeros (1000, 1);
```

creates the *V* variable of the required size and fills it with zeros. Every loop then results in replacing one element. This is a more efficient process. This is also an example of MATLAB warning, including suggested action available in 'warning strip' at the level of relevant script line in MATLAB Editor, and one of the reasons for using the MATLAB Editor for scripting.

First Steps: Algorithmization

From the previous simple example, one should understand the essential principles of working in MATLAB, particularly script development. However, the most difficult skill to learn is not to remember the many MATLAB inbuilt functions including their inputs and outputs, nor is it to use them appropriately for a given set of inputs, or combining them in elegant composite commands. The most difficult skill to learn is algorithm development. For the purpose of this book, the process of algorithm development can be split in three levels:

Level 1: Breaking a problem into a step-by-step sequence of smaller sub-problems whose sub-solutions can be combined in the overall solution at the end

Level 2: Consider all (or as many as) possible situations (input parameters variations, etc.) at each step within a given problem. A good program should be capable of dealing with 'everything possible' (even if it means only generating an error message describing the cause of the program crash)

Level 3: The highest level, which would involve optimization of calculation speed and use of computer memory

This is not a general scientific definition of the algorithm development process. Rather this is a description of what level of programming this book uses. Although Level 3 is considered and is sometimes crucial, most situations in this book deal with Levels 1 and 2. These are the levels necessary to solve a given problem (Level 1) using (ideally) a robust algorithm (Level 2). This means that although some applications may be comparable to similar commercial solutions in terms of quality of solution, calculation speed and memory handling is usually inferior. Again we emphasize that the major purpose of this book is to use a very basic level of programming to demonstrate the philosophy and complexity of some solutions in clinical radiotherapy physics.

The simple example above, calculating the total volume of one thousand cylinder shaped swimming pools, already required some decisions that could be considered algorithm development steps:

- Choosing the input data format, the *rh* variable
  - Having input data organized as a simple variable is usually a strong prerequisite for a quality solution. Extracting and organizing data from whatever is available in the real world may be much more difficult.

- Deciding to create a function for a single pair of *r* and *h* and calculating and recording the volume for each 'case' (swimming pool) from the *rh* input in the *V* vector and calculating the sum at the end…
  - Pre-allocating the *V* array with zeros can be considered as the simplest example of Level 3 program optimization.

Workspace Saving and Loading

There is no unique workflow one has to follow to work in MATLAB. Generally, one works around the inputs and operational variables created. Sometimes it may be useful to 'save workspace' to carry on later, mostly when working in the MATLAB Command Window instead of scripting in the editor. Workspace saving and loading later may ease a re-start point simply because there is no need to create all variables again.

'Workspace' means all variables currently defined. Using the *Save workspace* command/ icon in MATLAB toolstrip saves all variables in one **.mat* file. Alternatively use the

```
>> save filename
```

command in the MATLAB Command Window. Just as when using the icon, MATLAB generates the *filename.mat* file and saves it in the current working folder.

To (re-)open all variables saved in a **.mat* file, use either *Open* icon in the toolstrip, or enter

```
>> load filename.mat
```

or

```
>> load filename
```

command in MATLAB Command Window anytime. See MATLAB Help for more details and examples.

Variable Saving and Loading

Saving one or more selected variables is effectively equivalent to saving a sub-workspace. This is done using commands in MATLAB Command Window, for example, for saving two variables from the current workspace.

```
>> save fileName variableName1 variableName2
```

Alternatively, there is a functional expression for the same thing:

```
>> save ('fileName.mat', 'variableName1', 'variableName2')
```

Note this time using the *save* command requires entering filename and names of variables to save as a string, that is, with apostrophes. See MATLAB Help for more details and examples.

Path Setting/Adding

Setting or adding a path to let MATLAB know where to search for your own scripts, when you wish to run them either directly from MATLAB Command Window or as part of another script to execute, is the last essential practicality described in this section. By default, MATLAB uses folders created during installation (of MATLAB and toolboxes) to find all inbuilt functions. Work including user's scripts and variables should be saved in user defined folders. Mostly, they should not be a single folder, but rather a whole folder structure, for example, organizing problems, problem specific inputs, problem specific variables, problem specific scripts, or general scripts. Whenever you use a command linked to your own work, MATLAB searches within default paths and paths added by you so it is necessary to specify folders to search in. Files, scripts and data should be well organized in folders in order to avoid confusion while working. For example, one should be careful not to create two versions of a script saved under the same name in different folders.

MATLAB path can be controlled by the *Set path* icon in MATLAB toolstrip, or using the *addpath* command from the MATLAB Command Window, for example,

```
>> addpath D:\book_MATLABinRT\DraftWord\Chapter_1_MATLABessentials
```

DATA TYPES IN MATLAB

Numeric Types

The basic and default numerical data type is *double*. Variables of type *double* are stored with 64-bit precision. Using the *single* function, numbers can be converted from double precision to single type with only 32 bits, saving memory but with lower precision number representation. The difference is nicely presented by the example (MATLAB Help Online):

```
>> double (single (3.14) - 3.14)
ans =
   1.0490e-07
```

Numerical precision is important in some applications presented later in this book, particularly when searching the exact match of two numbers requiring, in theory, zero difference.

Other numeric types include, for example, *uint16* (unsigned integer with range from 0 to 65535) and *int16* (signed integer with range –32768 to 32767, reserving one bit for sign information). More information about numeric data types and mutual conversions can be found easily in MATLAB Help, for example, using the following command:

```
>> help datatypes
```

Data types and numbers format are important as many MATLAB inbuilt functions require a specific data type on input which may be different to another inbuilt function output so conversion between data types would be required.

Character and String Data

Next to numeric data it is common to also process characters and character strings: text data. Working with file and folder names to handle import or export data files provides a simple example.

From MATLAB Help: Introducing character strings:

S = 'Any Characters' creates a character array, or string. The string is actually a vector whose components are the numeric codes for the characters. The actual characters displayed depend on the character set encoding for a given font. The length of S is the number of characters (MATLAB Help Offline). Table 1.5 provides examples of a string variable definition.

Cell Arrays

Sometimes it is necessary to combine variables of different types and sizes. Cell array is a data type used in MATLAB to address such situations. Table 1.6 demonstrates the essentials of creating a cell array variable and addressing its components.

Structure Arrays

Similar to cell arrays, structure arrays allow containing variables of different data types and sizes, but it is also possible to organize data in categories, eg, fields with names. In a simplified way, structure arrays can be imagined as a kind of database. In the following example, there is data of similar structure for two patients which need to be organized in one structure array variable. There are few fields to form the required structure: patient *name*, date of birth (*DOB*), *hospital* and (type) of *therapy*. If therapy type is '*radio*', indicating radiotherapy, then treatment *dose* and number of *fractions* need be specified in

TABLE 1.5   Example of basic string variable definition and addressing

Command Window		Comments
>>	S = 'example string' S = example string	% create new variable, a character string
>>	whos Name Size Bytes Class Attr. S 1x14 28 char	% display the variable properties
>>	S(4:6) = 'MPL' S = exaMPLe string	% address three characters of the character array (string) and replace by another string (in this case replacing 'mpl' with 'MPL')

TABLE 1.6   Demonstrating essentials of cell type data

Command Window	Comments
>> `a = 3;`	% define numeric variable *a*
>> `b = 1:5;`	% define numeric vector variable *b*
>> `c = ones (4,5);`	% define numeric matrix variable *c*
>> `d = 'string variable';`	% define example string variable *d*
>> `eCell = {a,b,c,d};`	% define cell array variable *eCell* of size 1 × 4. Each cell contains one variable of different data type defined previously

>>   `aa = eCell(1)`
`aa =`
`[3]`
	% creating new *aa* variable by allocating the first *eCell* component. The *aa* is type cell

>>   `aaa = eCell{1}`
`aaa =`
`3`
	% creating new *aaa* variable by allocating the first *eCell* component. The *aaa* is (original, numeric) type *double*

>> `whos`	% list of all current variables, their size and types

Name	Size	Bytes	Class	Att's
a	1 × 1	8	double	
aa	1 × 1	120	cell	
aaa	1 × 1	8	double	
b	1 × 5	40	double	
c	4 × 5	60	double	
d	1 × 15	30	char	
eCell	1 × 4	686	cel	

respective fields too. Table 1.7 illustrates the building of an appropriate structure array using assignment statements, that is, entering data element-by-element. See MATLAB Help for alternative structure array building method (using *struct* function) and organizing database rather 'in planes' than element-by-element for better access to whole fields (when preferred).

For this book, structure array is a very important data type because much radiotherapy data imports into MATLAB as structures (see the section 'Structure Array' in Chapter 2).

BASIC OUTPUT AND EXPORT OPTIONS

Output format and user interactivity are essential for each practical application. Using MATLAB (or any other programming platform) for data processing requires careful choice of both form and format for both data input and output. Basic output data formats are numerical data, possibly organized in vectors or matrices or figures or multi-format data structures. For quality and clarity purposes, numerical data require clear labeling or headers. Figures need good titles, axes labeling, data legends, and so on. The choice of output format is naturally determined by application objectives: Just a record to be viewed by a user when needed? Processed data presentation for a user to make assessment and decisions? Or further processing in any other software product is required so the data format must be adjusted accordingly?

TABLE 1.7 Structure array building and elements access example

Command Window	Comments
>> A(1).Name = 'Smith John'	% define first patient's first field (*name*) value
>> A(2).Name = 'Evans Joseph'	% same for the second patient
>> A(1).DOB = '20-May-1956'	% define first patient's *DOB* as specific string
>> A(2).DOB =' 23-Jun-1966'	% same for the second patient
>> A(1).Hospital = 'Hospital 1'	% enter the first patient's *hospital* – string
>> A(2).Hospital = 'Hospital 2'	% same for the second patient – string
>> A(1).Therapy = 'Chemo'	% enter *therapy* type for the first patient – string
>> A(2).Therapy = 'Radio'	% same for the second patient – string
>> A(2).Therapy.Radio.Dose = 36.25	% enter *dose* for the second patient's radiotherapy
>> A(2).Therapy.Radio.Fractions = 5	% enter number of *fractions* for the second patient's radiotherapy

>>   A
A =
1 × 2 struct array with fields:
   Name
   DOB
   Hospital
   Therapy
	% created structure array variable *A* displayed (for more than one record only first field names, data type(s) and size is displayed)

>> A(1).Therapy ans = Chemo	% access the first patient's therapy type
>> A(2).Therapy ans = Radio: [1 × 1 struct]	% access the second patient's therapy type

>>   A(2).Therapy.Radio
ans =
   Dose: 36.2500
   Fractions: 5
	% access the first patient's radiotherapy details

>> A(2).Therapy.Radio.Dose ans = 36.2500	% access the first patient's radiotherapy dose details

When using MATLAB, essential output forms are the screen (Command Window and Workspace) and figures. Although figures can be saved manually and variable values copied over from the MATLAB environment, this is probably not the best approach for efficient applications of created programs. However, it can be efficiently used during the development phase.

In previous sections and Table 1.4, MATLAB *save* and *saveas* commands were introduced to demonstrate saving figures and workspace or individual variables. Although figures can be exported in standard figure formats, saving variables using *save* saves the data in MATLAB format (.*mat*) which does not support further processing without MATLAB. Therefore, MATLAB supports export in many common data formats. Examples of MATLAB inbuilt functions for data export are shown in Table 1.8.

TABLE 1.8 Some data export functions

MATLAB Function	Description
xlswrite	Write to Microsoft Excel spreadsheet file a two-dimensional numeric or character array or, if each cell contains a single element, a cell array.
csvwrite	Write matrix into a comma-separated value file. Cell arrays are not accepted for the input matrix.
dlmwrite	Write matrix into an ASCII delimited file. Cell arrays are not accepted for the input matrix.
dicomwrite	Write images as DICOM files. Optional parameters enable writing metadata (DICOM header) too.
xmlwrite	Serialize an XML Document Object Model node.

Note: See MATLAB Help for details.

The format specific data exports presented in Table 1.8 mostly require specific input which is not always convenient. For example, one needs to write a simple report file including both a rectangular data table and also other information forms, for example, the data header. This is typical for situations when MATLAB output needs to be expressed in a format specific to a third-party software for further processing; an approach very useful for independent software testing.

In the next section, we will demonstrate more flexible file printing using the *fprintf* function. Let's assume a third-party software requires a 2D dose matrix import as shown in Table 1.9.

In addition to the dose data table itself, there is much more information required such as X, Y coordinates for each row and column, data separator and data unit. Data may need be organized into header and body with start and end indicated, and so on.

TABLE 1.9 Example of a dose matrix data format required by third-party software (Web Supplement 1.1)

```
<exporttestascii>
<asciiheader>
Separator        [TAB]
Data Unit        Gy
Length Unit      mm
No. of           5
 Columns
No. of Rows      3
</asciiheader>
<asciibody>
X [mm]           0.0      7.5      15.0     22.5     30.0
Y [mm]
0.0              2.0000   2.0000   2.0000   2.0000   2.0000
7.5              2.0000   2.0000   2.0000   2.0000   2.0000
15.0             2.0000   2.0000   2.0000   2.0000   2.0000
</asciibody>
</exporttestascii>
```

Using the MATLAB *fprintf* inbuilt function let's create own function printing any rectangular dose matrix in MATLAB into a TAB separated text file corresponding to the sample format from Table 1.9. The next parameters required are the exported file name (string) and separation in both X (columns) and Y (rows) direction (1 × 2 vector). The script including a point-by-point explanation is shown in Table 1.10.

The function can be tested using the following simple command sequence in order to obtain the *testASCII.txt* file which, after importing in MS Excel (both TAB and colon delimiter on import), gives data as presented in Table 1.9. The data can be saved in MS Excel format for data import demonstration in the next section.

```
>> A = 2 * ones(3, 5);  %sample dose matrix
>> asciiExportTest ('testASCII.txt', A, [7.5,7.5]);   %calling the
function for specific parameters
```

The function example is also available in Web Supplement 1.1.

BASIC INPUT AND IMPORT OPTIONS

Unlike output options presented in the previous section, data import is generally less flexible. This is because it is the programmer's and end-user's preferences for exporting or presenting program outcome. This generally provides more options than when one is forced to deal with data formats produced (or required) by third party software. In radiotherapy (physics) there are many software applications allowing data export for further processing. On the other hand, a lot of data is not available for export at all, mainly for safety reasons and/or possible medical device status. Where data is available for export, it is a positive quality sign when more, and rather common formats are available (such as ASCII text, MS Excel, etc.). Sometimes even the copy -> paste function enables easy data transfer in some common format. Table 1.11 presents MATLAB basic import functions for the most common data formats.

Table 1.9 showed an example data format including both a numeric data table and additional text and numeric information. Table 1.10 presents the function exporting any rectangular dose matrix plus additional information in a text file of required format using the *fprintf* function. Let's now try to import the data from the created file *testASCII.txt* back in MATLAB.

From Table 1.11, the load function, for non-*mat* files, requires a rectangular table, so the only way to import the file using *load* is to modify the file in a text editor first: cut header and footer, replace the 'X[mm]' element with an arbitrary number, and remove line with 'Y[mm]'. Then both *load* and *dlmread* work:

```
>> result1 = load ('testASCIImod.txt')
>> result2 = dlmread ('testASCIImod.txt') %leaving delimiter to
auto detection
>> result3 = dlmread ('testASCIImod.txt', '\t') %including
delimiter specification
```

TABLE 1.10 Example script of a user-defined MATLAB function exporting a rectangular dose matrix in a desired output file format using *fprintf*

MATLAB Function	Description
```function [] = asciiExportTest(filename, doseMatrix, dataSpacing)```	% define function exporting a rectangular dose matrix in Gy as a TAB separated text file with some sample specific info. The matrix first column are respective Y-coordinates. X-coordinates will be printed as a separate line in the output. Note empty function output, not needed for objectives of such function.
```testFileID = fopen(filename, 'w');```	% opens a file with desired filename (see application example). Parameter 'w' means replacing existing file
```fprintf(testFileID,'<exporttestascii>\r\n');```	% Line 1: just a text <exporttestascii> as an example message for a third party software. Escape character \n means new line but combination with carriage return \r\n is stronger to move to a new line
```fprintf(testFileID,'\r\n');```	% empty line
```fprintf(testFileID,'<asciiheader>\r\n');```	% example text message for a third party software
```fprintf(testFileID,'Separator: [TAB]\r\n');```	% example message for a third party software – text: separation sign
```fprintf(testFileID,'Data Unit: Gy\r\n')```	% example message – text: data unit
```fprintf(testFileID,'Length Unit: mm\r\n')```	% example message – text: data unit
```fprintf(testFileID,'No. of Columns:%3.0f\r\n',size(doseMatrixInGy,2))```	% example message: data matrix size in columns. Number format: 3 characters including decimal point, 0 decimals
```fprintf(testFileID,'No. of Rows:%3.0f\r\n',size(doseMatrixInGy,1));```	% same as above for matrix rows
```fprintf(testFileID,'</asciiheader>\r\n');```	% example message for a third party software
```fprintf(testFileID,'<asciibody>\r\n');```	% example text message for a third party software
```Xscale=0: dataSpacing(2):(size(doseMatrixInGy,2)-1)* dataSpacing(2);```	% preparing X-coordinates (columns) vector: *dataSpacing*(2) mm spacing
```Yscale=0: dataSpacing(1):(size(doseMatrixInGy,1)-1)* dataSpacing(1);```	% same as above for the Y-coordinates
```doseMatrixScaleY=zeros(size(doseMatrix,1),size(doseMatrix,2)+1); doseMatrixScaleY(:,2:size(doseMatrixScaleY,2))=doseMatrix; doseMatrixScaleY(:,1)=Yscale';```	% expanding the dose matrix by the Y-coordinates column (1) to form composed dose and Y scale matrix for printing
```Vx='X[mm] \t'; for i=1:size(doseMatrixInGy,2)-1 Vx =[Vx,'%5.1f \t']; end Vx =[Vx,'%5.1f \t\r\n'];```	% preparing string defining number print format and separation for X-coordinates forming the first line of the data table. String elements are determined by dose matrix column size. First and last string elements are specific. \t means tab separation

(Continued)

TABLE 1.10 (CONTINUED)   Example script of a user-defined MATLAB function exporting a rectangular dose matrix in a desired output file format using *fprintf*

MATLAB Function	Description

```Vdose ='%5.1f\t';```
 ```for i=1:size(doseMatrixInGy,2)-1```
 ``` Vdose =[Vdose,'%7.4f\t'];```
 ```end```
 ```Vdose =[Vdose,'%7.4f\r\n'];```
	% analogical to above for dose matrix (column) elements

```fprintf(testFileID, Vx, Xscale);```	% printing X-coordinates vector *Xscale* using format given by *Vx*
```fprintf(testFileID,'Y[mm]\r\n');```	% next line indicating only that first column of the following table will be Y-coordinates
```fprintf(testFileID, Vdose,``` ```doseMatrixScaleY');```	% printing dose matrix elements including Y-coordinates specified in the first column using pre-defined *Vdose* format specification
```fprintf(testFileID,'</``` ```asciibody>\r\n');```	% example message for a third party software
```fprintf(testFileID,'</``` ```exporttestascii>\r\n');```	% example message for a third party software
```fclose(testFileID);```	% closing the file

TABLE 1.11   Some data import functions

MATLAB Function	Description
```RESULT = load(FILENAME)```	Load the variables from a MAT-file into a structure array, or data from an ASCII file into a double-precision array. ASCII files must contain a rectangular table of numbers, with an equal number of elements in each row. The file delimiter (character between each element in a row) can be a blank, comma, semicolon or tab
```RESULT = dlmread (FILENAME,DELIMITER)```	Read numeric data from the ASCII delimited file FILENAME using the delimiter DELIMITER. The result is returned in RESULT. Use '\t' to specify a tab
```RESULT = csvread(FILENAME)```	Read a comma separated value formatted file FILENAME. The result is returned in M. The file can only contain numeric values
```importdata```	Loads data from file. See demonstration in the text ...
```textscan```	Read formatted data from text file or string
```xlsread```	Read MS Excel file. See demonstration in the text ...
```dicomread```	Read DICOM image (and dose) ... see it applied in chapters that follow
```dicominfo```	Read metadata from DICOM message
```xmlread```	Read XML file and returns it as Document Object Model node

Note: See MATLAB Help for details.

With one line shorter, for example, line 3 (gray cells in Table 1.12), the file does not import using *load* but still it does using *dlmread* when 'missing elements' import as zeros.

Although the given examples show a few simple tricks on how to import using MATLAB functions requiring a certain format, the original file modification before import is clearly not an optimal way since some information is omitted (eg, units in the given example) and need to be added or considered manually. The *importdata* function is more efficient for such data.

TABLE 1.12   Modified text file from Table 1.9 to enable (partial) import using *load* and *dlmread*

9999	0	7.5	15	22.5	30	37.5	45
0	2	2	2	2	2	2	2
7.5	2	2	2	2	2	2	2
15	2	2	2	2	2	2	2
22.5	2	2	2	2	2	2	2
30	2	2	2	2	2	2	2

```
>> A = importdata ('testASCII.txt','\t');
```

import the example text file as a structure array *A* with two recognized components: *data* and *textdata*. There may be other structures recognized such as *colheaders* with appropriate function parameters specified (see MATLAB Help example) but that is not the case in here (from a format perspective), relatively complex example. Table 1.13 shows both components of the structure A as imported in MATLAB.

It is easy to process the *A.data* component, for example, separating dose data from the X positions, as it imports as a simple numeric matrix:

```
>> X = A.data (1,:); doseData = A.data (3:7,:);
```

To extract, for example, the *Y* positions it is a little trickier for it imports as cells within the *textdata* component (gray area items 13 to 17 in Table 1.13):

```
>> for i = 13:17 Y(i) = str2num (A.textdata {i}); end
>> Y (Y >0)

ans =
   7.5000  15.0000  22.5000  30.0000
```

The presented example of a text file import, simple in content, but relatively more complex in format (different from a rectangular data table), demonstrates practical issues one can face

TABLE 1.13   Two recognized structure components following import of sample *testASCII.txt* file using *importdata*

>> A.textdata		>> A.data						
ans =		ans =						
'<exporttestascii>'	-continued-	0.00	7.50	15.00	22.50	30.00	37.50	45.00
''	'X [mm] '	NaN	NaN	NaN	NaN	NaN	NaN	NaN
'<asciiheader>'	'Y [mm] '	2.00	2.00	2.00	2.00	2.00	2.00	2.00
'Separator: [TAB]'	'0.0'	2.00	2.00	2.00	2.00	2.00	2.00	2.00
'Data Unit: Gy'	'7.5'	2.00	2.00	2.00	2.00	2.00	2.00	2.00
'Length Unit: mm'	'15.0'	2.00	2.00	2.00	2.00	2.00	2.00	2.00
'No. of Columns: 7'	'22.5'	2.00	2.00	2.00	2.00	2.00	2.00	2.00
'No. of Rows: 5'	'30.0'							
'</asciiheader>'	'</asciibody>'							
'<asciibody>'	'</exporttestascii>'							

when dealing with data import. Non-standard file and data format is specific and there is no 100% universal and reliable import solution, so it is a programmer's job to find a solution which is robust enough to cover all, or as many as possible, of the potential variations.

The import of standard file formats can be demonstrated with the same test data used so far (Table 1.9), but imported in MS Excel and saved in *xls(x)* format as suggested at the end of the previous section. A spreadsheet file import in MATLAB can be tested by using the *xlsread* inbuilt function:

```
>> [NUM, TXT, RAW] = xlsread ('testASCII.xls');
```

Import the sample spreadsheet as three components:

NUM: Numeric matrix of elements identified as numbers, all other elements import as NaN

TXT: Cell array of elements identified as text

RAW: Cell array with all elements, all other elements such as empty row or empty cells that are complementary to rectangular data table import as NaN

Here is an example of addressing a cell with a supposed numeric content including string to number conversion:

```
>> str2num (RAW{15, 4})
ans =
     2
```

A demonstration of importing other data formats using functions listed in Table 1.11 can be found either in MATLAB Help (*textscan*) or in the next sections when they are used to solve specific problems (*dicomread, xmlread*).

MANAGING FILES AND FOLDERS

Now that we know how to generate some export files (including figures) within a script, it is clear that sometimes it is necessary to organize files in folders. This includes file moving and copying, navigating the folder structure, listing folder contents and creating new folders. Table 1.14 presents the basic relevant commands and their application.

A SIMPLE CASE STUDY

The following example is the closest so far to the scope of this book. Using a very simple example from clinical practice, it presents building an algorithm to solve a given problem and demonstrates practical circumstances to be considered in order to make the solution accurate and robust.

TABLE 1.14   Essential files and folder operations

Command (Example)	Description
`cd`	% displays current working directory path
`cd folderName`	% change current working directory to *folderName*
`cd (folderName)`	% same as above when *folderName* is stored in a string
`cd..`	% moves to the directory above the current one
`dir`	% lists current directory content
`dcmFiles = dir ('*.dcm');`	% returns the structure array of all *dcm* files in current folder. The structure fields are: *name, date, bytes, isdir, datenum ...*
`file3name = dcmFiles (3).` `name`	% ... returns the name of the 3rd *dcm* file as a string

`eval (['cd`
`',foldersNames(i).`
`name])`
	% provided *foldersNames* variable contains the *dir* structure of folders in current directory as per above, this expression allows dynamic navigating in folders by evaluating appropriate string thus converting it into executable command

`copyfile`
`('sourceFileName','`
`targetFileName')`
	% copies a source file to a target file specified by their names as strings. Source and target can be both specified as either an absolute pathname or a pathname relative to the current directory

`delete fileNameToDelete`	% deletes a file from disk
`delete (fileNameToDelete)`	% deletes file with file name stored in a string
`movefile`	% move a file, application syntax analogical to *copyfile* above
`rmdir`	% remove empty directory, application syntax analogical to *delete* (for files) above
`mkdir`	% create directory, syntax analogical to *delete* (for files) above

Note: See MATLAB Help for more details and application examples.

Problem Definition

Determine radiation field size from a radiation dose profile. Sounds trivial? Well …

Problem Specification

Question number one is whether the problem definition as stated above is detailed enough to design a solution. This example sounds trivial for a radiotherapy physicist: based on established guidelines and practices, the radiation field size is defined as the respective dimension of 50% isodose line (eg, IAEA 2005) where 100% is meant the central beam axis (CAX) dose. Let's assume a symmetric and non-small field for this moment (see Figure 1.3 for an example). Input dose profiles supposedly measured at a reference distance to radiation source, reference depth, and reference medium for all these parameters may impact the parameter value. However, there are questions that require answers to ensure the volume and quality of the information is enough before looking for a solution:

- What will be the input data format? Will it always be the same or must some variation be considered? For example, will positions always be given in millimeters? How about dose units? Gy, cGy, relative percentage (range 0–100%), fraction (range 0–1), all of these?

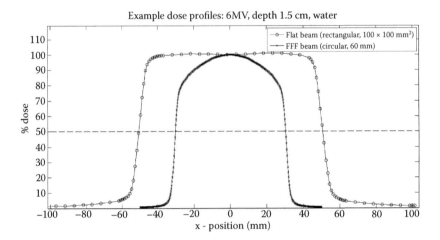

FIGURE 1.3   Example of two 6-MV photon beam dose profiles at 1.5 cm depth of water. The dashed line indicates a level of 50% of the CAX dose where field size parameter is defined.

- Can we always expect equidistant data spacing? Or, for example, should higher resolution be expected in profile shoulder regions (field penumbra)?

- What level of accuracy is required?

- Should the solution cover both flat and flattening-filter-free (FFF) radiation beams (assuming photon beams only)?

- Would electron or proton (or other modality) beams need be considered too?

- What exactly is the objective for all this? Is it to determine field sizes for beam output dependence? Is it to measure one of the field parameters for stability checks as part of a QA program? Is there another purpose? Sometimes it might be useful to think even 'outside the box', eg, if the beam output dependence is the objective, could perhaps the field area instead of the field 'diameter' be a better parameter? In this case probably not, since from a physics perspective this is about the radiation scatter contribution to dose on beam axis, so the projected distance of field edge from the axis is the most important factor. How about if the parameter for QA stability check is the objective? Would it be better to approach the problem from a different perspective? For example, rather than *field size* measuring a *collimator jaw position*? Although very similar technically, might not such an alternative better address what is actually needed?

- What is the desired output format? A parameter value printed on-screen? The result exported as a text file? Dose profile plot to include in output with calculated field size displayed in graph title or data legend? Or should the solution be in the form of a function presuming the output will be used as part of some other, bigger, program, depending on objective?

But there are even more important questions …

- Should asymmetric fields be included/considered in the solution? This is a crucial question and directly related to the main objective, then the definition of the 50% isodose line does not work as beam CAX may not be covered by dose profile available.

- If CAX normalization is required, is the CAX position clearly indicated in the profile positions vector, eg, corresponding to zero position?

- Considering the main objective, are reference conditions for input data a necessary prerequisite? If so, does the input data format support or include relevant information to verify before parameter calculation starts?

- If, for example, the input dose profile was measured at other than reference distance and if we know about it, should the solution use the information and correct the result (in this case using the similar triangles principle)? Or, should it simply generate an error message and stop?

From the presented case and possible variations of problem specification, one can already imagine what a programmer needs to deal with and that many important decisions need be made in order to design a quality solution even for problem that might sound trivial in the beginning.

Algorithmization

Of all the questions listed above, let's define something like an operational framework to design an appropriate algorithm. Let's create a function calculating *full width at half maximum* (FWHM) of a given dose profile on input as a reasonable parameter describing field size. Using a profile maximum, or rather a robust maximum (discussed later in this chapter), it is not strictly required that CAX dose is covered by profile data. This makes the approach more flexible, yet without the significant impact on the solution for FFF beams CAX dose should be the maximum, and for flat beams an over-compensated dip in profile around CAX is small, so the error associated with not following the definition strictly is also negligible (see the 'Testing' section later in this chapter). The FWHM approach also includes radiation background response by definition so that is one aspect less to worry about. See Figure 1.3 for a graphic demonstration. Let's assume the input dose profile as a $[x, D(x)]$ array where positions x may not be equidistant. And let's make the decision that the only function output required is the calculated field size value. Since the accuracy was not specified in the problem statement, let's decide that the required precision of the calculated parameter is 0.2 mm. Let's consider that all other aspects discussed in the previous section are dealt with outside the defined operational framework. The decisions just described in this section should be sufficient enough to develop an appropriate algorithm described in the following scheme.

Algorithm Scheme

Function Syntax
fwhm (doseProfile)
Function Input/Output

> Input
>> *doseProfile* Radiation dose profile $[x, D(x)]$, $N \times 2$ numeric array; column 1 = x positions in millimeters, column 2 = $D(x)$ dose at position x (any dose unit)
>
> Output
>> *fwhm* Single number on the function output

Algorithm Steps

- Find dose profile maximum
- Find dose profile minimum (radiation background)
- Interpolate radiation profile from spacing on input to equidistant 0.1 mm spacing: $x \to xi$
- Find all positions xi of the interpolated profile with dose above *(Dmax – Dmin)/2*
- Determine FWHM as the difference in the *ix* positions between the last and the first point found in the previous step

Alternative Options

a. There is always some noise present in real data. When the normalization constant is used and based on a single point such as global maximum or minimum, there is always a chance it is an outlier. This may not be a big problem when the data can be inspected visually beforehand. If this is not case, then an alternative, more robust solution is preferred. In the given example, both profile maximum and minimum can be determined, eg, using histogram-based statistics.

b. In the given example, the CAX dose is usually used for profile normalization in order to determine the field size. If this is required, and it is reasonable to expect CAX to be identical with the field center, it is possible to consider using the default algorithm (based on extreme point) initially to determine the profile center as the average position of all profile points with dose above, eg, 50%, and then use the dose of the central point to renormalize the profile to derive the 'true', 50% level and complete field size calculation the same way as in the original algorithm.

c. Another option to determine the field size based on the CAX dose, is that radiation profile positions are typically symmetric around zero indicating CAX based on some reference applied during the measurement setup. In such a case, it is simple to find the profile point at zero position and use this to (re)normalize the profile before and carry on with the original algorithm.

Scripting

Following the previous algorithm scheme, write a script (in MATLAB Editor). An example function script is shown in Table 1.15. The script, together with sample dose profiles to test it, are available in Web Supplement 1.2.

TABLE 1.15   Function *fwhm* script to calculate the FWHM of a dose profile

MATLAB Function	Description
`function y = fwhm (doseProfile)`	% define function calculating the FWHM from the radiation dose profile given as a N × 2 array of positions in the first and dose in the second column, respectively
`Dmax = max (doseProfile (:, 2));`	% determine dose profile maximum
`Dmin = min (doseProfile (:, 2));`	% determine dose profile minimum
`xi = doseProfile (1, 1):0.1:doseProfile(end,1);`	% define new positions for interpolated profile *xi* keeping the same range as original profile but with equidistant spacing 0.1 mm
`Di = interp1 (doseProfile(:, 1), doseProfile(:, 2), xi);`	% calculate 1D interpolation for dose *Di* for the original positions, original dose and new positions *xi* input parameters
`xiDoseAbove50 = xi (find (Di > (Dmax -Dmin)/2));`	% find new positions *xi* with new dose above 50% of difference between *Dmax* and *Dmin*
`y = xiDoseAbove50(end) -xiDoseAbove50(1);`	% assign the result with the difference between the edge positions of selected points (FWHM)

The following command sequence in the MATLAB Command Window can then be used to test the function using two sample profiles (Web Supplement 1.2).

```
>> load sampleProfiles
>> fwhm (circProfile60)
ans =
      60.8000
>> fwhm (squareProfile100)
ans =
      101.5000
```

Testing

Once a script is producing accurate results using sample data one needs to move to another phase: testing and fine tuning. Some aspects such as intrinsic accuracy in the given example are determined by the algorithm parameters. For example, with a 0.1 mm fixed resolution for the interpolated dose profile there is a maximum +/- 0.2 mm uncertainty in the result *due to this factor* which is acceptable for most clinical applications. However, there are always other aspects that are not covered directly by the algorithm and that need to be considered and, based on potential impact on patient outcomes, assessed whether they should be considered explicitly. One category of such aspects is well described by the question: *What can go wrong?* The following points show some examples for the above case of field size:

- What if the profile does not have the expected shape, that is, slow rise – sharp rise – flat area (flat beams) or peak (FFF) – sharp fall – slow fall? If this is relevant, then some automatic extension should be considered in order to test the profile shape before or during analysis. One can think of a condition such that a vector of profile points above

50% level, that is, step 4 in the algorithm scheme above, has no gaps (eliminating unrealistic multiple peak profile) and also none of the profile edge points are included (eliminating no peak, eg, half profile data only is another possible situation). Then only the hypothetical situation detected would be the case of multiple peaks of which all smaller peaks have maxima below 50% of the highest peak. Alternatively, a profile gradient curve can be calculated in order to determine the number of local extremes and make an assessment based on that. Obviously, thinking that deep is unnecessary in such a simple application, but it can serve as well as an example for possible consideration.

- Effect of normalization. Regardless of official definition, dose profile normalization based on point maximum, robust maximum or CAX dose were all considered in the description of the algorithm and alternative options. It is a simple lab exercise to test sensitivity of the result to the data normalization approach. Results for two sample profiles from Figure 1.3 for simulated normalization levels are shown in Table 1.16. Considering that for FFF beams the CAX and position of dose maximum should be identical, and that for flat beams the overcompensated dip in profile around CAX is around 3% maximum, any associated variation of the result would not be dramatic. However, the point here is that the author of the case study application considered this aspect during testing. Given the data in Table 1.16 and considering the required accuracy of the application too, one could conclude that the issue of normalization is not dramatic. So for practical applications it is reasonable to stick with the original algorithm based on a single point maximum and minimum, provided that data quality is *reasonable* unless, of course, the CAX normalization is strictly required.

- The last example relates to the algorithm alternative, option a) replacing point D_{max} with a more robust normalization value, for example, a center of the last non-zero dose bin counting *reasonable* number of points, for example, 2%. Number of points to calculate the average depend on the total length of the profile which is certainly a variable parameter. There may be situations, though extreme, that such 2% selection of total profile may in fact represent 10 or even 20% of the nominal field size (eg, nominal field size 5 mm, profile spacing 0.1 mm, profile length 5× the nominal field size = 250 points, 2% of the total profile gives 5 points representing 10% of nominal field size). Depending on the case, this factor may mean that too many points are used to increase the statistical weight of a given parameter since the selection covers points that are not equivalent from the physics nature of the data. A solution might be controlling the selection parameter based on nominal field size or absolute number of

TABLE 1.16 Example of sensitivity of the field size parameter to profile normalization

	Simulated Error Factor				
Dose profile/nominal field size	0.94	0.97	1.00	1.03	1.06
FFF/circular/60 mm	61.0	61.0	60.8	60.6	60.4
flat/square/100 mm	101.9	101.7	101.5	101.1	100.9

Note: Simulated error factor represents a 6%, 3% and 0% difference of normalization dose in both directions to the default (profile *Dmax* value).

points or a combination of both. Having simulated input data covering as many data variations as possible is an efficient approach to study the behavior of a given application. For understandable reasons, the testing phase requires inputing either endpoint user or an application specialist with practical knowledge of the relevant field.

Documentation

Although teaching how to produce commercial level applications to be used in the clinical environment certainly is not the objective of this book, it should be mentioned that like with any other object of human work, establishing and following some quality standards is useful, especially in terms of efficacy. Appropriate documentation certainly contributes to keeping work products organized and in order. This book deals with selected case problems and their solutions using scripts developed inhouse, and as such, belong to the category of bespoke software. Using bespoke software in the clinical environment and medical device status for software have been an important and heavily discussed topic recently among relevant regulators including, for example, the European Commission (eg, EC 2017/745) and U.S. Food and Drug Administration (eg, SAMD 2016). The general purpose of this book is teaching, so applications/scripts developed to demonstrate a given solution should not be used routinely in clinics. They do not require meeting formal requirements; however, if it happens that this book contributes to someone to develop their own script to solve a particular, whatever minor, problem (which is what the author hopes for), then the following can be recommended in terms of documentation:

- Each script should contain a header. Style of comments in MATLAB scripts can serve this purpose well. The header should contain essential information in order to control the script application status, owner, date of last modification, version indicator, brief description of what the script does, summary of input and output data, input data requirement, etc.

- Scripts should contain comments to help others understand and remind the writer about the rationale behind the script.

- In case a particular script is used for some well defined and/or periodic activity (eg, as part of QA program), general standards for equipment commissioning should be followed. This should include a commissioning report and work instructions. The report should include information about what tests were applied and what results were achieved. The script and related documentation should be covered by a general quality system which would guarantee formal quality requirements such as periodic revision, version control, ownerships and personal responsibilities, etc. Work instruction should provide application use safety guidelines to each defined user group. The work instructions document can also include troubleshooting and/or actions to take based on the results obtained. The document can take the form of a user manual.

REFERENCES

EC 2017/745. (2017). Medical devices regulation. https://publications.europa.eu/en/publication-detail/-/publication/83bdc18f-315d-11e7-9412-01aa75ed71a1/language-en. Accessed January 2018.

IAEA. (2005). *Radiation oncology physics: A handbook for teachers and students.* Podgorsak EB, Printed by the IAEA in Austria, July 2005, ISBN 92-0-107304-6Q2.

MATLAB Help Offline R2013a. (2013). MATLAB, The MathWorks, Inc., Massachusetts, USA.

MATLAB Help Online. (2017). The MathWorks, Inc. https://www.mathworks.com/help/matlab/matlab_prog/floating-point-numbers.html. Accessed May 11, 2017.

MATLAB. (2017). The MathWorks, Inc. https://www.mathworks.com. Accessed May 11, 2017.

MS Excel: Microsoft Office, Microsoft Corporation. (2017). http://microsoft.com. Accessed May, 11, 2017.

Software as Medical Device (SAMD): Clinical Evaluation. (2016). https://www.fda.gov/downloads/medicaldevices/deviceregulationandguidance/guidancedocuments/ucm524904.pdf. Accessed May 11, 2017.

CHAPTER 2

Radiotherapy Physics Related Data Types and Basic Operations

LEARNING OUTCOMES

LO 2.1 Perform basic operations with 1D data commonly used in radiotherapy physics including dose profiles and depth dose curves: display, mirroring, symmetric averaging, smoothing, interpolating and matching

LO 2.2 Perform basic operations with dose-volume histograms: Converting between differential and integral form, display and calculating major statistical parameters

LO 2.3 Perform basic operations with 2D data commonly used in radiotherapy including images and dose planes: Interpolation/resize, smoothing, rotating, display, matching and defining regions-of-interest

LO 2.4 Explain basic operations on 2D to 3D/4D data examples: 3D/4D CT

LO 2.5 Perform operations on MATLAB cell and structure array data types with data commonly used in radiotherapy: Organizing 4D CT series, and organizing and addressing selected DICOM metadata from a CT series

LO 2.6 Perform operations with string variables: Address and modify parts of strings

INTRODUCTION

This chapter provides examples of operations on data typical to radiotherapy including introduction to operations such as display and basic manipulation of such data. These basic operations will be used later to design more complex algorithms in order to solve selected problems. The data examples are presented based on dimensionality and/or data types that are used in MATLAB. Regarding data dimensionality, note the following working convention: 1D-array/line (often referred to as 'vector'), 2D-array/plane (often referred to as 2D-matrix), 3D-array/volume (often referred to as 3D-matrix).

SCALARS

There are countless examples of scalar variables in radiotherapy: air temperature needed for dose measurements, ionization current, mean target/organ dose, and so on.

1D DATA

One-dimensional data can be expressed using a common format [x, y(x)] such as [x-coordinate, dose at x] for the cross-plane dose profiles, [depth, dose] for depth dose curve, [time, respiratory amplitude] for respiratory cycle record, [dose, volume] for dose-volume histogram (DVH), or [control point number, MLC-leaf position] for MLC control point sequence, and so on. Values may or may not be given in equidistant intervals or increments. For example, for depth dose curves it is common to increase the sampling frequency close to the water phantom surface where curve shape changes rapidly. Similar considerations may apply to the beam penumbra region of dose profiles. In general, sampling frequency should be a function of the rates of change of the quantity with the independent variable. Let's use sample dose profiles and DVH data – imported and saved in the MATLAB format already (Web Supplement 2.1) – to demonstrate some basic operations starting with data display.

Graphical Display

Beam Scans

The sample dose profile (Web Supplement 2.1) is an example of 1D data. However, the best way of handling such data within MATLAB is to use a two-dimensional array with 1D position (x-coordinate in millimeters) specified in the first, and corresponding dose in the second column, respectively.

The sequence of commands in Table 1.4 and Figure 1.2 (see Chapter 1) demonstrated basic graph plotting and labeling by using the *sin(x)* function. The sample dose profile analogy is Table 2.1 (command sequence) and resulting Figure 1.3.

TABLE 2.1   Basic 1D data display

Command Window	Comments
>> load sampleProfiles	% load prepared sample relative dose profiles from *sampleProfiles.mat* file
>> plot (profile60 (:, 1), profile60 (:, 2), 'bx-')	% create new (1st) figure object and plot relative dose vs. position data (solid blue line with crosses indicating data points)
>> hold on	% hold data displayed in current figure object (1) to display additional data in the same figure
>> plot (profile100 (:, 1), profile100 (:, 2), 'ro-')	% plot additional data; next sample dose profile
>> xlabel ('x - position [mm]')	% add x-axis (horizontal) label
>> ylabel ('%Dose')	% add y-axis (vertical) label
>> title ('Example dose profiles: 6MV, depth 1.5 cm, water')	% add figure title
>> legend ('FFF beam (circular, 60mm)', 'Flat beam (rectangular, 100 x 100 mm2)')	% add displayed data legend
>> hold off	% release data hold for current figure object

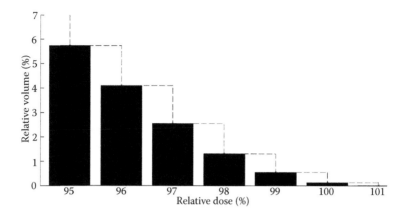

FIGURE 2.1   Zoomed sample DVH plot using *plot* (x), *bar* and *stairs* (—) functions.

Dose Volume Histogram

The *dose volume histogram* (DVH) is used routinely in radiotherapy to express, in a statistical way, the dose distribution coverage of *volumes-of-interest* (VOIs) such as targets and critical structures (or organs-at-risk, OAR). In the most common integral (cumulative) form, and with dose and volume both relative, a DVH point $[D_i, V_i]$ represents the fraction V_i of a given VOI receiving a dose greater than or equal to D_i. This is expressed as the fraction of the normalization dose (typically absorbed doses are normalized to prescribed dose or dose maximum). Using DVH sample data (Web Supplement 2.1), one can demonstrate basic DVH plotting using *plot, bar* or *stairs* MATLAB inbuilt functions. The following command sequence produces the DVH plots shown in Figure 2.1.

```
>> load sampleDVHs
>> hold on
>> plot (relDose, relVolume1,'x'); bar (relDose, relVolume1);
>> stairs (relDose, relVolume1, '—')
>> xlabel ('Relative dose [%]'); ylabel ('Relative volume [%]')
```

Interpolating Dose Profiles

Data interpolation is a key operation for many applications of 1D data, for example, resampling/change of sampling interval (sometimes referred to as 'spatial resolution'). In Table 1.15 (see Chapter 1) we applied the MATLAB function *interp1* already when interpolating dose profiles in order to determine the radiation field size parameter. More examples of data resampling are demonstrated using another sample dose profile available in Web Supplement 2.1 and the following command sequence:

```
>> load sampleProfiles
>> x = profile100 (:, 1); D = profile100 (:, 2);
>> whos x D
Name   Size          Bytes         Class
   D    80 × 1         640           double
   x    80 × 1         640           double
```

```
>> interpProfile_A = interp1 (x, D, [min(x):.5: max(x)]');
>> interpProfile_B = interp1 (x, D, [min(x):.5: 1.2*max(x)]');
>> interpProfile_B1 = interp1 (x, D, [min(x):.5: 1.2*max(x)]',
   'linear', 'extrap');
```

Note that the original position 1D array x and (transposed) 1D array specifying interpolated points both form one column arrays. Also note that specifying the interpolation method and the difference in handling new points requested outside the range of the original position 1D-array – B and B1 request extrapolation outside original range – $1.2\ x$ $max(x)$. The difference between B and B1 is the extrapolation request necessary to extrapolate data beyond the original range.

It should be understood that 'interpolation' using the *interp1* function technically means 'resampling using interpolation'. For the given example, based on dose given at original positions with variable spacing (5 mm/1 mm), new data points were calculated for new positions (equispaced positions with 0.5 mm spacing) using linear interpolation of the original data points (Figure 2.2).

Should someone need a quick view of spacing variation then, for the example case with positions defined by the 1D array x, plotting a 1D-array of differentials can do the job:

```
>> plot (diff (x))
```

See MATLAB Help for more information on extrapolation methods, and so on.

Matching Dose Profiles

Comparing two dose profiles visually is common in radiotherapy physics applications for purposes of commissioning or QA. There are two operations to facilitate such a task:

1. Dose profile normalization is usually carried out with three options:

   - Dose at CAX defined as reference (zero) position: CAX_0

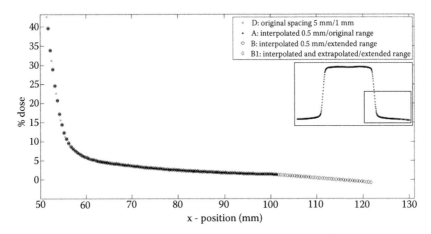

FIGURE 2.2 (Zoomed) Example dose profile resampling using (linear piecewise) interpolation and extrapolation.

- Dose at CAX defined as beam profile center: CAX_C

- Dose (more robust) maximum: D_{max}

2. Profiles match to reference point defined as one of four options again:

- CAX_0, CAX_C

- One of two beam edges (*edgeL/edgeR*)

Which of these options to use depends on the application and/or agreed standards. For example, to assess the stability of radiation beam shape for a fixed collimator, it is reasonable to match beam profiles based on the beam center ($CAXc$). To assess accuracy of collimator jaw opening relative to collimator rotation axis by comparing profiles acquired at two opposite collimator angles, using a scanner reference (eg, CAX_0) is clearly a better option. To demonstrate matching profiles, one should be familiar with finding possible matching reference first.

Finding CAX_0

When the reference position is already included in the dose profile position 1D-aray, the problem is solved. If the reference coordinate is not present, then it is possible to resample the data using interpolation with a new position 1D-array, including the desired reference coordinate (eg, zero).

Finding CAX_C

The beam profile center has been described already (see the section 'A Simple Case Study' in Chapter 1) as the arithmetic mean of the edge positions of data points with dose above given level (eg, 50% of normalization value). If profile data sampling is insufficient relative to the desired accuracy, the profile can be interpolated. In the very likely situation that the calculated CAX_C position is not available directly within the original profile positions, then resampling with new positions, including calculated CAX_C is an option.

Finding edgeR/edgeL

It is uncommon to match two dose profiles using either beam edge, however, to find a position of the left or right profile edge at a given dose level (eg, 50% of normalization value again) is technically analogous to finding the previous CAX_C, only instead of the mean one would use the respective edge position directly.

Once the reference positions for both dose profiles are found, then completing the profiles match is simple by shifting either position 1D-array by respective reference position and making a new reference position equal for both dose profiles, for example, zero. This results in dose profiles with equivalent position 1D-arrays with identical position coordinate for either reference point, that is,

```
>> x_new = x_orig - xRef
```

Where *x_new* and *x_orig* are original and new position 1D-array, respectively, and *xRef* is a reference point position (eg, *CAXc*).

However, range and spacing can still be different. This is not a problem for graphic presentation (see Figure 1.3), but if identical position 1D-arrays are required for matched profiles, resampling both (shifted) profiles with the desired universal position 1D-array is an option. In such cases, a new common position 1D-array can be defined to cover the overlapping range, that is, from *max(min(x1,x2)* to *min(max(x1), max(x2))* where *x1* and *x2* are respective position 1D-arrays. To make sure the reference position (eg, zero) is within the new 1D-array one can consider the following construction (equidistant from the reference zero in both sides of the respective edge of the overlapping range):

```
[fliplr(-desiredStep:-desiredStep:max(min(x1,x2)),0:desiredStep:min
(max(x1),max(x2))]
```

Table 2.2 demonstrates the sequence of MATLAB commands to match two dose profiles from sample profiles data in Web Supplement 2.1. The data contains two dose profiles with a simulated CAX_0 position error of −0.5 mm and +0.3 mm, respectively. The task is to match two dose profiles using the beam center as the reference (CAX_C). The result is shown in Figure 2.3.

Sometimes it might be useful to match two 1D data sets using the reference 1D-array indices. The following function matches two 1D-arrays (*A, B*) with respective reference indices (*refA, refB*). The result is identical in size to the first 1D-array (*A*).

```
function newB = match1Dto1st (A, refA, B, refB)
newB = zeros(size(A));
helpVar = B(refB-min(refA,refB)+1:refB+min(length(A)-
refA,length(B)-refB));%help variable
newB(refA-min(refA,refB)+1:refA+min(length(A)-refA,length(B)-refB)) =
helpVar;
```

Note that using this approach for dose profile matching requires identical spacing of both position 1D-arrays.

Symmetric Averaging Dose Profile

Typical operations with dose profiles reflect the purpose and method used in their acquisition. Typically, 'off-axis ratio' or 'off-center ratio' (OCR) tables are required for beam models in treatment planning systems. Dose profiles are usually measured along two axes perpendicular to the central beam axis (cross-plane/in-plane) with a range well beyond the radiation field edge. One way to obtain OCR from a dose profile is *averaging* doses at symmetric positions. The sequence of commands demonstrating manual *symmetric averaging* dose profile using the sample data is presented in Table 2.3. Assumptions are only that the zero position (not necessarily included in the position 1D-array) indicates center of profile symmetry, otherwise position 1D-array needs adjustment (shift) for this to be the case.

TABLE 2.2. Matching two dose profiles using beam center reference (CAX_C)

Command Window	Comments
`>> load sampleProfiles`	% load prepared sample relative dose profiles from *sampleProfiles.mat* file
`>> p1 = profile60Asymm0p5mm;`	% duplicate loaded variables to shorten code length below
`>> p2 = profile60Asymm0p3mm;`	
`>> whos p1 p2`	% show profiles parameters: different range, generally different spacing, position
`   Name    Size        Bytes       Class`	1D-array (column 1), dose 1D-array (column 2)
`   p1      501 × 2     8016        double`	
`   p2      484 × 2     7744        double`	
`>> p1i(:,1)=min(p1(:,1)):.1:ma`	% interpolate/resample profile 1: same range, fixed 0.1 mm spacing. Again, column 1 for new positions, column 2 for new dose
`   x(p1(:,1))'; p1i(:,2)=interp1(p1(:,1)`	
`   ,p1(:,2),p1i(:,1));`	
`>> p2i(:,1)=min(p2(:,1)):.1:ma`	% ... analogically for profile 2
`   x(p2(:,1))'; p2i(:,2)=interp1(p2(:,1)`	
`   ,p2(:,2),p2i(:,1));`	
`>> p1i50 = p1i(find(p1i(:,2)>0.5*(max(p1`	% from the interpolated profile p1i select positions (column 1) with dose (column 2) above 50% of dose range
`   i(:,2))+min(p1i (:,2)))),1);`	
`>> p2i50 = p2i(find(p2i(:,2)>0.5*(max(p2`	% ... analogically for profile 2
`   i(:,2))+min(p2i (:,2)))),1);`	
`>> xP1_CAX = (p1i50 (end) + p1i50 (1))/2`	% calculate mid-position of the interpolated profile points with dose above 50% — CAX_C
`   xP1_CAX =`	
`     -0.5000`	
`>> xP2_CAX = (p2i50 (end) + p2i50 (1))/2`	% ... analogically for profile 2
`   xP2_CAX =`	
`      0.3000`	
`>> plot(p1(:,1),p1(:,2),'bx')`	% plot original profiles 1 and 2, then the same with respective CAX_C shift
`   hold on`	
`   plot(p2(:,1),p2(:,2),'ro')`	
`   plot(p1(:,1)-xP1_CAX,p1(:,2),'kx')`	
`   plot(p2(:,1)-xP2_CAX,p2(:,2),'ko')`	
`   hold off`	

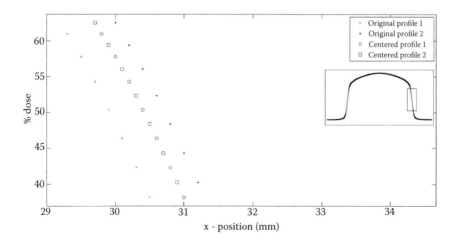

FIGURE 2.3 Matching two dose profiles using beam center reference (CAX_C).

TABLE 2.3   From full to symmetric average half profile: Manual

Command Window	Comments
`>>    load sampleProfiles`	% load prepared sample relative dose profiles from *sampleProfiles.mat* file
`>>    x = profile100(:,1);` `>>    D = profile100(:,2);`	% duplicate loaded variables to shorten code length below
`>>    plot(x, D, 'xb'); hold on`	% plot relative dose vs. position data (blue crosses)

`>>    xi = [fliplr(-1:-1:-min(abs([m`
`      in(x),max(x)]))),0,1:1:min(abs`
`>>    ([min(x),max(x)]))]';`
`>>    Di = interp1(x, D, xi);`
	% resample dose profile with equidistant 1 mm spacing in symmetric range given by original absolute position extreme closer to zero

`>>    indexPositionsNonNeg = find` `      (xi >= 0);`	% create/find 1D-array of indices of non-negative positions
`>>    indexPositionsNonPos = find` `      (xi <= 0);`	% create/find 1D-array of indices of non-positive positions
`>>    indexPositionsNonPosFlipUD =` `      flipud(indexPositionsNonPos);`	% flip the 1D-array upside down (up-down)

`>>    meanHalfProfile = mean([Di(ind`
`      exPositionsNonNeg)',`
`      Di(indexPositionsNonPosFli`
`      pUD)']);`
	% create new 1D-array of the same length as the mean of two relative doses at the opposite positions 1D-array of respective indices is used to address relative doses at matching positions. Note 1D-arrays dimensionality for correct mean calculation

`>>    plot(xi(indexPositionsNonNeg),` `      meanHalfProfile,'or')`	% plot relative dose vs. position data – only half profile with non-negative positions (red line), mean calculated using *mean* command. Note transposed dimensionality and yet correct display

Mirroring Dose Profile

The opposite operation of *mirroring* to obtain a perfectly symmetric profile from an available half-profile can be demonstrated by using the following command sequence and the corresponding sample data available in Web Supplement 2.1. The only assumption is that, again, zero position represents center of symmetry for the assumed mirroring operation.

```
>> load sampleProfiles
>> x = halfProfile(:, 1); D = halfProfile(:, 2);
>> positionsFull = [flipud(-x); x];
>> doseFull = [flipud(D); D];
>> plot (x, D, 'ok'); hold on; plot (positionsFull, doseFull,
   'xk'); hold off
>> title ('Mirrored half dose profile'); xlabel ('x [mm]'); ylabel
   ('Relative dose [%]');
>> legend ('Original half profile', 'Mirrored full profile')
```

Note that if a half-profile contains zero position explicitly, then the resulting full profile contains two identical points at zero position which is not a problem for display but may be a problem for further processing. The following sequence demonstrates identical operation with selecting only indices of positive positions to be used for mirroring operation:

```
>> load sampleHalfProfiles
>> positionsFull = [flipud (-x (x >0)); x];% potential zero position
   point from original x only
>> doseFull = [flipud (D (x >0)); D];
```

... display same as above.

Compared with mathematically similar operations in the symmetric averaging above, there is more use of composite commands without the *find* function in the last example.

Smoothing a Dose Profile

Smoothing is common for 1D-data such as a (percentage) depth dose curve. Here it will be demonstrated by using beam profiles. In order to try smoothing there needs to be a noisy profile available. Using simple and perfect sample data it is possible to generate random noise and corrupt a profile to test the basic smoothing method. Figure 2.4 shows the original profile, the profile with added noise and two smoothed profiles in order to demonstrate the effect of the *medfilt1* MATLAB inbuilt function which calculates the median of the defined number of points around a given data point:

```
>> load sampleProfiles
>> whos profile60
   Name            Size        Bytes       Class
   profile60       501 x 2     8016        double
>> myNoise =
   randn (size (profile60 (:, 2)));
>> noisyProfile = 0.2 * myNoise + profile60 (:, 2);
>> smoothProfile_N5 = medfilt1(noisyProfile, 5);
>> smoothProfile_N10 = medfilt1(noisyProfile,10);
>> plot (profile60 (:, 1), profile60 (:, 2), 'bx-'); hold on;
```

... plot other three data sets

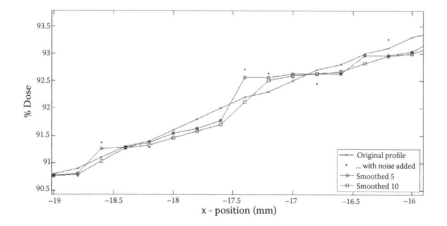

FIGURE 2.4   Original profile, with added noise and smoothed profiles using the *medfilt1* function with 5 and 10 neighbours.

DVH: Differential and Integral Format

Conversion among various forms of DVH is one of the essentials. In general, a DVH can take any combination of the following forms:

- integral (cumulative) vs. differential (frequency)
- absolute vs. relative dose
- absolute vs. relative volume

As any 1D data, a DVH can be seen as two biased 1D-arrays [dose] and [volume], so to convert absolute dose/volume to relative and vice versa it is a trivial operation of multiplying or dividing a 1D-array by normalization constant, for example, for the dose:

```
>>  relDose1D = absDose1D/NormDoseConstant *100; % to obtain
    percentage values
```

Conversion between integral and differential form is more interesting. Let's design two MATLAB functions *int2diff* and *diff2int* by converting one DVH type to the other. It will keep the relative or absolute form of dose and volume used on input. Before writing any script we need to understand the relationship between the two types of histograms. Differential DVH shows relative or absolute counts of voxels (volume) within a given VOI with dose within a given dose bin. The integral (cumulative) form represents volume of a given VOI receiving dose greater than or equal to a given dose value. The relationship between the two types is expressed in:

$$V_i^{int} = \sum_{k=i}^{N} V_k^{diff} \text{ and } V_i^{diff} = V_i^{int} - V_{i+1}^{int}$$

where N is number of dose bins and i and k relate to given dose bin.

Let the *diffH* be the input variable (differential DVH) in the form of $N \times 2$ array with the first column containing doses (bins) and the second column with associated volumes. Then the *diff2int* conversion function can be defined as shown in Table 2.4:

TABLE 2.4   Function converting differential DVH to integral DVH

MATLAB Editor	Comments

```function intH = diff2int (diffH)```
```intH (:, 1) = diffH (:, 1);```
```for i = 1:size (intH, 1)```
``` intH (i, 2) = sum (diffH (i:size```
``` (diffH,1), 2));```
```end```
	1. Initialize function for a differential DVH input
	2. Integral DVH's dose 1D-array remains the same
	3. For each dose bin D_i of the integral DVH ...
	... calculate the sum of differential DVH volumes
	$[V_i, V_{i+p...}, V_{last}]$

The output dose 1D-array (first column) is identical. Units for both dose and volume can be whichever (Gy, %dose, cm³, mm³, %volume), the output will have the same units. The opposite conversion function *int2diff* requires integral histogram *intH* on input and the script is analogous:

```
function diffH = int2diff (intH)
diffH(:, 1) = intH (:, 1);
for i = 1:size (intH, 1) - 1
        diffH (i, 2)=intH (i, 2) - intH (i+1, 2);
end
```

Once we have the functions they need to be tested. In this case there are few objectives for the test evaluation, for example, the output of one function used as the input for another must be equal to the original input data, and total volume in the differential histogram should be equal to the first bin volume of the integral histogram. The following command sequence performs such test using DVH sample data available in Web Supplement 2.1:

```
>> load sampleDVHs
>> whos
   Name             Size          Bytes         Class
   relDose          101 × 1       808           double
   relVolume1       101 × 1       808           double
   relVolume2       101 × 1       808           double
>> intH(:, 1) = relDose; intH (:, 2) = relVolume1;
>> diffH = int2diff (intH); intHre = diff2int (diffH);
>> plot (intH (:, 1), intH (:,2), 'xb')
>> hold on
>> plot (diffH (:, 1), diffH (:, 2), '.r');  plot (intHre(:, 1),
   intHre (:, 2), '-k')
>> xlabel ('Relative dose [%]'); ylabel ('Relative volume [%]')
>> legend ('original intH','int2diff (intH)', 'diff2int (int2diff
   (intH))')
```

Figure 2.5 demonstrates the test rest: integral histogram of the differential histogram calculated from the original integral histogram equals to the original integral histogram.

DVH Statistics

The next basic operation demonstrated in this section will be DVH-based statistics *minimum dose*, *mean dose,* and *maximum dose*. It is important to understand that determining these basic VOI statistics from a DVH is not 100% equivalent to doing the same based on original population of voxel doses belonging to a given VOI. This is simply because DVH washes out fine differences in voxel doses within a bin, effectively by replacing the distribution of voxels with dose within a given bin by its median.

The minimum dose from a differential DVH (*diffH*) is the first dose bin with non-zero count of voxels or volume. Analogously, maximum dose from a differential DVH (*diffH*) is

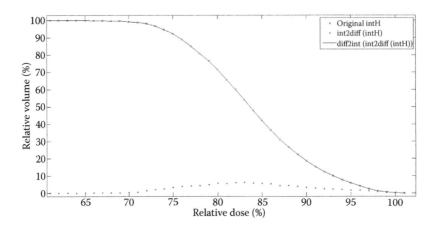

FIGURE 2.5 Integral to differential and back DVH conversion test.

the last dose bin with non-zero volume. Following a DVH format from the example above then:

```
nonZeroBins = find (diffH, 2) > 0    % find dose bins with non-zero
                                       volume
minDVHdose = diffH (nonZeroBins (1), 1)   % the first dose bin is
                                            the minimum
maxDVHdose = diffH (nonZeroBins (end), 1) % the last dose bin is
                                            the maximum
```

Whether a DVH-based dose minimum/maximum equals either global extreme voxel dose or is slightly different depends on how dose bins are constructed and labeled. In principle there are two options to apply: bin edges or bin centers. See MATLAB Help for *hist* and *histc* inbuilt functions.

The mean dose to a given VOI can be calculated from a differential DVH easily as a weighted sum, that is, partial volumes serve as weight factors for a given dose bin:

```
function DVHmean = meanDVHdose (diffH)
DVHmean = 0;
totalV = sum (diffH (:, 2));
for j = 1:size (diffH, 1)
          DVHmean = DVHmean + diffH(j, 1) * diffH (j, 2)/totalV;
end
```

The differential histogram input format is the same for previous functions, and dose units are the same as the input. Mean target dose from the integral DVH sample can be then calculated in the MATLAB command window using the two new functions:

```
>> load sampleDVHs
>> intH (:, 1) = relDose; intH (:, 2) = relVolume1;
```

TABLE 2.5   Example of PTV and bladder basic dose statistics (Gy) based on a full population of voxels forming a given VOI, and based on DVH-derived statistics for different number of dose bins used in a differential histogram. *N* indicates the number of voxels in respective bins. Dose bins are labeled with their centers

PTV (45143 voxels)				
minDose =	41.5184	*maxDose* =	47.0898	*meanDose* =	46.1994
bins	*minDVHdose*	*N(1)*	*maxDVHdose*	*N(end)*	*meanDVHdose*
10	41.797	2	46.8113	4332	46.2178
50	41.5741	1	47.0341	15	46.1992
100	41.5463	1	47.062	4	46.1995
200	41.5323	1	47.0759	1	46.1995
Bladder (14497 voxels)				
minDose =	8.2887	*maxDose* =	46.5258	*meanDose* =	21.7535
bins	*minDVHdose*	*N(1)*	*maxDVHdose*	*N(end)*	*meanDVHdose*
10	10.2005	2644	44.6139	801	21.7313
50	8.671	186	46.1434	160	21.7539
100	8.4798	56	46.3346	58	21.7547
200	8.3843	15	46.4302	18	21.7538

```
>> meanDVHdose (int2diff (intH))
ans =
   83.4410
```

The minimum and maximum doses are important dose distribution parameters in radiotherapy. Although rare there may be situations when basing these values on a single voxel with an outlier dose value is misleading and a more robust approach is preferred. For example, ICRU 83 guidelines recommend using $D_{2\%}$ from an integral DVH to be reported as representative of target dose maximum. Table 2.5 shows an example of DVH statistics from a clinical dose distribution for PTV and bladder as organ at risk. PTV dose is rather homogeneous, bladder dose is the opposite; both result from the standard treatment planning strategy. The table shows statistics based on the full population of voxels forming a given VOI and statistics calculated from the respective DVH using the methods previously described. The point is to demonstrate the number of voxels a given DVH statistic is based on and also that parameters values are quite robust with respect to a (reasonable) number of bins.

2D DATA

All two-dimensional data can be expressed using a common format $[x, y, z(x,y)]$ such as $[x\text{-coordinate}, y\text{-coordinate}, D(x,y)]$ for 2D dose distribution, or $[x\text{-coordinate}, y\text{-coordinate}, $ CT-number] for a CT transverse image which, from a data type perspective, is equivalent to any planar (eg, X-ray) image. Values may or may not be given in equidistant intervals or increments. For example, CT resolution in the transverse plane is typically identical in both directions by default. But for reconstructed sagittal or coronal plane, this may not be the case because one dimension is determined by the CT slice thickness, or better *z*-spacing, which is generally different from transverse pixel size. Let's consider a sample CT transverse slice imported and saved in MATLAB format (Web Supplement 2.3). (How to import image data into MATLAB is explained in Chapter 3.

Graphical Display

We will demonstrate essential operations typical for such 2D data, starting with the data display. The command sequence in Table 2.6 results in the image seen in Figure 2.6.

Alternatively, *imagesc* can be used instead of *imshow* when display as 'rather 2D-array' while indicating 2D-array coordinates on the axes is preferred. Try it yourself and see MATLAB Help for details.

Image overlays are common in radiotherapy. Although more sophisticated tools are typically used for multimodality anatomic data display, the basic *imshow, imagesc* and *contour* commands as presented is usually sufficient for displaying dose, isodoses and/or isodoses overlaid on CT anatomy images, or for simple manual image fusion as demonstrated in Chapter 5. Note that although the demonstrated image looks like an anatomy image with specific *region of interest*, ROIs (bones in green, body and cavity contours in red), at the moment this is merely a display. How to create an actual ROI is explained later in this chapter (see the section '2D Data/Regions of Interest' later in this chapter).

TABLE 2.6   Basic 2D data display

Command Window	Comments
>> load sampleCTax	% load sample CT transverse slice (2D)
>> whos	% list all the variables (1) in the current workspace
Name Size Bytes Class sampleCTax 512 × 512 2097152 double	
>> imshow (sampleCTax, [])	% scale the data and display the 2D 2D-array as an image. Empty 2D-array [] means full range of pixel values is used for 64-bit grayscale imaging window
>> colorbar	% display colorbar in the figure
>> impixelinfo	% turn on pixel information tool in current image
>> imshow (sampleCTax, [1500 2500])	% re-display the image now with imaging window specified by restricting 2D-array values to display in the 64-bit colormap (bone contrast)
>> hold on	% hold current figure
>> contour (sampleCTax,[1500:500:2500] ,'g')	% display the same data (2D-array) using *contour* in range focused on bone in green
>> contour (sampleCTax,[500:10:600],'r')	% same as above, now with focus on tissue-air interface in red
>> caxis ([900 1400])	% keeping the image just adjusting image window, now with bones and air emphasized adjusted to soft tissue again

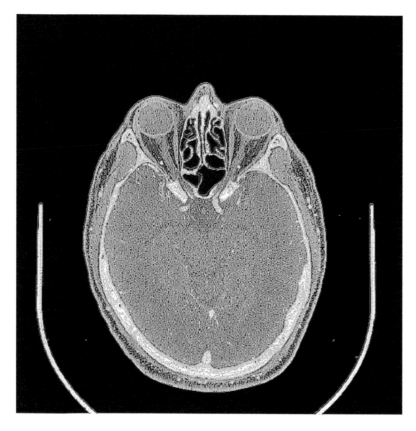

FIGURE 2.6. Example of transverse CT image with two contour sets (plots) added: red indicates air interfaces, while green indicates bone interfaces.

Interpolation and Resizing

The first essential operation with 2D data in radiotherapy presented here is changing image size (pixel size), and therefore spatial resolution which is frequently applied to both medical images and 2D dose distributions. 2D-array CT image pixel size is typically given by the dimensions of the *field-of-view* (FOV) in order to cover the desired anatomy divided by the reconstruction 2D-array size. The manually defined FOV is then reconstructed using, for example, 512 × 512 2D-array of pixels which means that unless the FOV is fixed, images from different patients or different images from the CT series of a particular patient have generally different pixel size. MATLAB has two inbuilt functions to change image size easily, *imresize* and *interp2*, however, it is important to understand what exactly change of size or resolution means. The *imresize* function works with a grayscale, RGB, or binary image. Mathematically, these are all 2D-arrays, just as a RGB image consists of three 2D-arrays representing each color channel (layer), and a binary image a 2D-array where values are ones or zeros. In terms of size or pixel size, MATLAB knows only the number of rows and columns forming a given 2D-array). A 2D data resize in radiotherapy most often means adjusting different but known pixel sizes (or, pixel-dose spacing in case of dose distribution

2D-arrays). To match pixel sizes of two images or two 2D-planar dose distributions one uses the ratio of pixel spacing values (Table 2.7).

There are differences in outcomes even while keeping the internal interpolation method parameter the same for both functions ('*bilinear*'). There is a slight geometrical shift of an object relative to image center when *interp2* is used. The *imresize* keeps objects aligned and also reduces noise. On other hand, in contrast to *interp2*, the *imresize* changes image depth by calculating image histograms using the command sequence shown in Table 2.8 and illustrated in Figure 2.7. In radiotherapy imaging and physics there are situations where either geometry, minimum impact on element values, or both are important so the approach should be chosen carefully. For example, unwanted smoothing associated with element values may not be acceptable for resizing dose distributions. However, the *imresize* function reproduces the original histogram when the '*nearest*' interpolation method is used as the function parameter. In conclusion, a function together with parameters must be chosen carefully with regard to the objective.

The MATLAB inbuilt function *hist* calculates a histogram. For images (2D-matrix) it calculates down the columns. Therefore, this is a straightforward command for 100 bins:

```
>> [N, X] = hist (sampleCTax, 100);
```

TABLE 2.7 Image resize using *imresize* and *interp2*

Command Window	Comments
>> `load sampleCTax`	% load sample CT transverse slice (2D)
>> `whos`	% list all the variables (1) in current workspace
`Name` `Size` `Bytes` `Class`	
`sampleCTax` `512 x 512` `2097152` `double`	

>>   `testImresize256=`
       `imresize (sampleCTax, 0.5,`
       `'bilinear');`
	% image resize using *imresize*, downsampling by factor of 2, preferred interpolation method

>> `testImresize256_nearest = imresize` `(sampleCTax, 0.5, 'nearest');`	
>> `[Xorig, Yorig] = meshgrid (1:512,` `1:512);`	% define original data grid for 2D interpolation
>> `[Xnew, Ynew] = meshgrid (1:2:512,` `1:2:512);`	% define new data grid for 2D interpolation: downsampling by factor of 2

>>   `testInterp256=`
       `interp2 (Xorig, Yorig, sampleCTax,`
       `Xnew, Ynew, 'bilinear');`
	% calculate 2D interpolation with preferred interpolation method

>>   `testImresize1024 =`
       `imresize (sampleCTax, [1024,1024],`
       `'bilinear');`
	% image resize using *imresize*, upsampling by factor of 2, preferred interpolation method

>> `[Xnew, Ynew] = meshgrid (0.5:0.5:512,` `0.5:0.5:512);`	% define new data grid for 2D interpolation: upsampling by factor of 2

>>   `testInterp1024 =`
       `interp2 (Xorig, Yorig, sampleCTax,`
       `Xnew, Ynew, 'bilinear');`
	% calculate 2D interpolation with preferred interpolation method

TABLE 2.8   Image resize: Comparing *imresize* and *interp2* using histograms

Command Window (Continued from Table 2.7)	Comments
`>>    [N,X] = hist(sampleCTax(:), 100);` `       hold on`	% calculate histogram of the sample image using 100 bins in range of the data values. *N* stores counts, *X* stores 1D-array of bin centers; image hold
`>>    plot (X, histc(sampleCTax(:),` `       X)/512/512*100, 'ko')`	% calculate and plot whole image histogram. Histogram counts are normalized by image size in pixels. Histogram calculated using *histc* with fixed bin centers (*X*)
`>>    plot (X, histc(testImresize256(:),` `       X)/256/256*100,'ks')`	% calculate and plot image histogram for bin centers *X* for the original image resized using the *imresize*
`>>    plot (X, histc(testInterp256(:),` `       X)/256/256*100, 'kd')`	% same as above for the image resized using *interp2*
`>>    plot (X, histc(testImresize256_` `       nearest(:), X)/256/256*100,'k.')`	

`>>    xlabel ('CT number')`
`       ylabel ('Relative counts [%]')`
`       legend('orig 512 × 512', 'imresize`
`       256 × 256', 'interp2 256 × 256')`
	% display axes labels and data legend

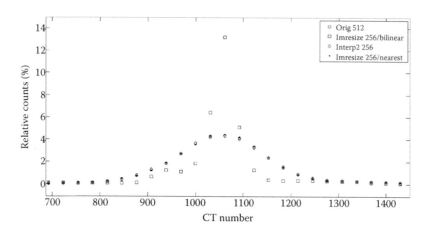

FIGURE 2.7.   Change of image (data) depth demonstrated using 2D histograms as products of *imresize* and *interp2*.

which results in *N*, a 2D-array of counts for individual columns. To obtain whole 2D-matrix counts, *N* needs to be summed by columns:

```
>> N1 = sum(N');
```

An easier way, also used in the example above, is to expand the source 2D-array so 2D data is represented 'unwrapped' as 1D data:

```
>> [N2, X2] = hist (sampleCTax (:), 100);
```

This results in identical (just transposed) bin centers in 1D-array, that is:

```
>> max(abs(X1-X'))
ans =
     0
```

and, the same 1D-array of counts:

```
>> max(abs(N2-N1))
ans =
     0
```

Filtering

The next basic image operation is filtering. Image smoothing is not used often directly in radiotherapy postprocessing, rather it forms part of the image acquisition process. For example, CT image protocols should be optimized for image noise vs. patient dose; this may include image smoothing. An example of MATLAB image filtering is the *medfilt2* function:

```
>> load sampleCTax
>> figure; imshow (sampleCTax, [900 1600])
>> figure; imshow (medfilt2 (sampleCTax), [900 1600])
```

See MATLAB Help for more information about 2D filtering.

Rotations, Mirroring, Zoom

Rotations, mirroring, and zoom are frequently required when working with radiotherapy images and dose planes. The MATLAB commands in Table 2.9 demonstrate these basic operations on sample image data.

TABLE 2.9 Basic rotation and mirroring

Command Window		Comments
>>	load sampleCTax	% load sample CT image 2D-array
>>	figure; imshow (sampleCTax, [])	% display sample image in new figure object
>>	figure; imshow (imrotate (sampleCTax, 45), [])	% rotate sample image by 45° and display it as a new figure. Rotated image is larger in size to contain all parts of rotated image
>>	figure; imshow(imrotate (sampleCTax, 45, 'crop'), [])	% rotate sample image by 45° and display it as a new figure. Rotated image has the same size as original by cropping to fit
>>	figure; imshow (flipud (sampleCTax), [])	% sample image flipped up-down
>>	figure; imshow (fliplr (sampleCTax), [])	% sample image flipped left-right

Image zoom is usually applied when figure viewing and possible interaction (eg, ROI or reference point definition), is required. For these applications zoom can be controlled intuitively by appropriate icons in the MATLAB figure toolbar. Alternatively, the MATLAB inbuilt function *zoom* can be used, mostly to activate special features of a general zoom application. Rather than zooming, applications presented in this book work with ROIs which is the next basic operation to demonstrate for 2D data.

Regions of Interest

As mentioned before, the most common 2D data in radiotherapy are anatomy images, or rather 2D slices of a 3D image series, and dose maps. Depending on application, there are ways and means to define the ROI. The first category is direct selection of 2D-array indices regardless of element value such as rectangle, circle, ellipse, or polygon. The second category is characteristic by ROI definition using a graphical drawing. And the last category can be considered elements selection by their value. Table 2.10 demonstrates the three categories described using different approaches and tools.

For practical applications each ROI can be characterized by a binary mask, a 2D-array of the same size as the original 2D data with ones in elements inside given ROI and zeros otherwise. Having such mask makes it easy to address or select relevant elements in the original data by using, for example, MATLAB function *find*. Let's have an original data *origData* and the same size binary *roiMask* with ones defining specific ROI. Then data elements (pixels) forming a given ROI in the original data can be addressed and processed easily by using the commands shown in Table 2.11.

TABLE 2.10   ROI data extraction from original 2D data using (ROI) mask

Command Window	Comments
`load sampleCTax` `>>`	% load sample CT image 2D-array

`>>   rectROImask =`
`zeros(size(sampleCTax);`
`rectROImask (340:450,340:450) = 1;`
`figure; imshow (rectROImask);`
`figure; imshow (sampleCTax`
`(340:450, 340:450), [])`
	% rectangular ROI mask; same size as the image with zeros everywhere except selected rectangular region filled with ones

`>>   E = [1 0.2 0.4 0.2 0.2 30];`
`>>   elipseROImask = phantom(E, 512);`
`>>   figure; imshow(elipseROImask)`
	% ellipse ROI mask; specify intensity, two semi-axes, center position, and angle

`>>   polyROImask = roipoly (sampleCTax`
`[340 340 450 450], [340 450 450`
`>>   340]);`
`figure; imshow (polyROImask)`
	% polygonal ROI mask: known vertexes (in pixel coordinates)

`>>   polyROImask1 = roipoly`
`(sampleCTax/max(sampleCTax (:)));`
`>>   figure; imshow (polyROImask1)`
	% polygonal ROI mask: vertexes defined using graphical interface and mouse; double-click to close the contour

`>>   boneROI = sampleCTax > 1600;` `>>   figure; imshow (boneROI)`	% ROI mask defined by element value selection criteria: all pixels with CT value over 1600

TABLE 2.11. ROI data extraction from original 2D data using (ROI) mask

Command Window	Comments
`>>` `roiLinIndices = find (roiMask == 1)`	% find linear indices of elements with value of 1 in the mask 2D-array
`>>` `roiVals = origData (roiLinIndices)`	% create 1D-array of original data values for elements forming the ROI defined by the mask
`>>` `[roiRows, roiCols] = ind2sub (size (origData), roiLinIndices)`	% convert linear indices to 2D-array row and column subscripts
`>>` `[roiRows, roiCols] = find(roiMask == 1)`	% alternative to same as above
`>>` `roiAllData = [roiRows,roiCols, roiVals]`	% organize all ROI elements in one N × 3 2D-array with their original 2D-array subscripts and values for further processing

Matching 2D Data

In an earlier section (1D Data/'Matching Dose Profiles') we demonstrated matching two 1D dose profiles $[x, D(x)]$ by adjusting positional coordinates x and data resampling. Matching using reference 1D-array indices was introduced as an alternative option. The most common multi-dimensional radiotherapy data this book deals with is stored and/or exported in DICOM format. As will be discussed in Chapter 3, MATLAB inbuilt functions enable quick import using a single command. The data is imported as a multi-dimensional array with row, column, or slice spacing specified as external information from data headers. Although reference point coordinates are usually extracted in Cartesian coordinates, very often it is easier and more natural to work with multi-dimensional arrays (eg, dose 2D-array) rather than with multiplets such as $[x, y, D(x, y)]$. These are why matching two 2D-arrays is demonstrated using reference 2D-array indices.

The most common application is matching a dose 2D-array with the same location anatomy (eg, CT) image to display isodose distribution which is demonstrated in detail in Chapter 3. Let's have two rectangular 2D-arrays of generally different sizes with a common reference point used for registration and specified by row and column indices. There are at least two options how to decide on the output format:

- Two new registered rectangular 2D-arrays of the same size corresponding to the overlapping region between the two original 2D-arrays

- One new rectangular 2D-array, the second input 2D-array registered to the first with zeros outside the overlapping region

Both options are demonstrated in Figure 2.8 and can be handled by simple functions (*match2D* and *match2Dto1st*) available in Web Supplement 2.4. The script is a 2D analogy to *match1Dto1st* function described previously ('Matching Dose Profiles'). The following command sequence demonstrates the *match2Dto1st* function in both directions on sample 2D-arrays from Figure 2.8.

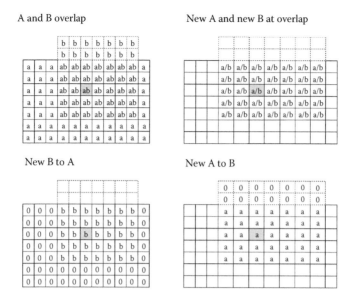

FIGURE 2.8   Outcome options (New A/B) of two overlapping 2D-arrays (A and B) registration.

```
>> A = ones (7, 11); refA = [3, 6]; B = 2*ones (7, 7); refB =
   [5, 3];
>> figure; imagesc (match2Dto1st (A, refA, B, refB)); colormap
   (gray)
>> figure; imagesc (match2Dto1st (B, refB, A, refA)); colormap
   (gray)
```

3D DATA

A common format for 3D data is an array $[x, y, z, f(x,y,z)]$ such as $[x$-coordinate, y-coordinate, z-coordinate, $D(x, y, z)]$ for 3D dose distribution, but obviously a more frequent form is a 3D-array of dose values where spatial location is given by array coordinates. The next examples of 3D data can be considered dynamic 2D images $[x, y, t, f(x, y, t)]$ with imaging time changes of a given quantity in a plane or projection.

Same as with 2D data, values may or may not be given in equidistant intervals or increments, but both the CT and dose in transverse plane typically have the same spacing. Spacing along the longitudinal axis is often given by slice thickness from an underlying 3D anatomy image. Let's have a sample CT series imported and saved in MATLAB format already (Web Supplement 2.3). How to import DICOM data in MATLAB is discussed in Chapter 3.

Graphical Display

In radiotherapy, displaying 3D images is mostly equivalent to displaying 2D images as 3D rendering is not used in practice as often as a 2D display of major anatomical planes. Basic displaying of CT data using a sample CT series is presented in Table 2.12. From a math perspective, the sample CT series is a 3D-array of CT numbers.

TABLE 2.12. Basic display of major anatomical planes of a 3D CT

Command Window	Comments
`>> load CT`	% load sample CT series saved in *mat* format
`>> whos`	% display all variables from current workspace and their main parameters; a 3D-array of CT numbers
Name Size Bytes Class CT 512 × 512 × 51 106954752 double	

`>> figure;`
`   imagesc (CT(:, :, 26)); colormap`
`   (gray)`
	% display transverse slice 26 of 51 (all rows and columns, fixed slice) in a new (gray) figure

`>> axis equal`	% equalizes increment size on both image axes

`>> ImgCor (:, :) = CT (256, :, :);`
`   figure`
`   imagesc (ImgCor)`
`   colormap (gray)`
	% select central coronal plane from the 3D series and displays it in a new figure. Note equalizing axes makes no sense now since CT pixel size and slice thickness are different anyway

`>> figure`
`   imagesc (squeeze (CT (256, :, :))`
`   colormap (gray)`
	% alternative for central coronal plane; this time using *squeeze* to handle 3D to 2D transition

`>> whos`	% display all variables from current workspace and their main parameters

Name   Size   Bytes   Class
    CT   512 × 512 × 51 106954752 double
   ImgCor 512 × 51 208896 double	

Interpolation and Resizing

Essentially, the processing of 3D data for radiotherapy applications is similar to 2D data, that is, (image) resize, interpolation, matching, and histogram calculation.

Similar to 2D there are two basic options for 3D data resize using either the *imresize* or the *interp3* MATLAB functions. The CT series sample array of dimensions $512 \times 512 \times 51$ has pixel spacing of 0.5059×0.5059 mm^2 in the transverse plane and a slice thickness (z-spacing) 4 mm. Let's resize the data into homogeneous isotropic resolution of $1 \times 1 \times 1$ mm^3 using 3D interpolation:

```
>> [Xorig, Yorig, Zorig] = meshgrid (1:512, 1:512, 1:51);
>> [Xnew, Ynew, Znew] = meshgrid (1:1/0.5059:512, 1:1/0.5059:512,
   1:1/4:51);
>> newCT = interp3 (Xorig, Yorig, Zorig, CT, Xnew, Ynew, Znew);
```

The resulting *newCT* matrix has dimensions $259 \times 259 \times 201$ with the required voxel size ratio. Major planes can be displayed analogously to the original CT as shown in Table 2.12.

Matching 3D Data

Matching two 3D-arrays is analogous to the 2D case presented in the previous section including optional preferences for outputs. The two simple functions (*match3D* and *match3Dto1st*), three-dimensional analogs to *match1Dto1st* presented previously in this chapter (see the section '1D Data/Interpolating Dose Profiles' earlier in this chapter), matching two 3D-arrays of any dimensions based on reference voxel coordinates are available in Web Supplement 2.5.

Rotations and Mirroring

Mirroring, or flipping a 3D-array can be done efficiently using the inbuilt function *flipdim (X, dim)* returning array *X* flipped along specified dimension *dim*. For a 3D-array, *dim* can be 1, 2 or 3, for rows, columns or slices, respectively).

An easy way to rotate a 3D-array along a specified dimension and central voxel in MATLAB is using the inbuilt function *imrotate* applied to individual planes (transverse, coronal, or sagittal). For example, rotating the sample CT model by 45° about the central voxel in the sagittal plane (fixed columns) can be achieved by the following command sequence:

```
>> load CTmodel
>> for col = 1:size (CTmodel,2)
       newCT (:,col,:) = reshape (imrotate (squeeze (CTmodel
              (:,col,:)), 45, 'crop'), size
              (CTmodel,1), 1, size (CTmodel,3));
   end
>> figure; imagesc (squeeze (newCT (:, 256,:))); colormap(gray)
```

Use the *squeeze* and *reshape* inbuilt functions to adjust object dimensions for *imrotate* and *imagesc* operations. Note the *'crop'* parameter to maintain the source image size.

Regions of Interest

The definition of ROIs is also analogous to the 2D situation. They can be defined directly by the range of array indices for cuboid shaped ROIs or by any other mathematical criteria, for example, a spherical ROI is formed by voxels with a given 3D distance from a reference voxel. ROIs can also be defined by selecting specific element values. Alternatively, the *phantom* function can be used too in principle to construct a 3D mask consisting of ellipses slice by slice, but this is rather impractical. More useful in practice are polygonal ROIs that, slice-by-slice, form a 3D mask. This is the same as for the 2D situation slice-specific polygons that can be either defined by known vertexes given as array indices, or they can be defined manually by slice-by-slice 'contouring'. In both situations, the *roipoly* function can be used efficiently.

Histogram

In the section above entitled '2D – Interpolation and Resizing', we calculated histograms for a whole image represented by a 2D array. The easiest way to calculate a histogram from a 3D array is to reorganize 3D data into a 1D column and use the default 1D histogram function *hist* again:

```
>> load CT
>> CToneColumn = CT (:);
>> [binCounts, binCentres] = hist (CToneColumn, 100);
>> figure; bar (binCentres, binCounts)
```

The MATLAB *reshape* function reorganizes 1D column data into a multi-dimensional array based on given dimensions:

```
>> CTcolumnTo3D = reshape (CToneColumn, size(CT));
```

In practice, however, histograms are usually calculated from ROIs typically defined by 3D masks of generally shaped objects. In that case it is efficient to use the linear indexing of the *find* function as demonstrated in the following example where the ROI is bone tissue with a CT number above 1600:

```
>> bone3Dmask = zeros(size(CT)); bone3Dmask (find (CT>1600)) = 1;
>> CTboneFromMask = CT (find (bone3Dmask == 1));
>> [binCounts, binCentres] = hist (CTboneFromMask, 100);
>> figure; bar (binCentres, binCounts)
```

Naturally, for ROIs defined directly by element values, the histogram can be calculated directly, that is, without use of a mask:

```
>> CTbone = CT (find (CT > 1600));
>> [binCounts, binCentres] = hist (CTbone, 100);
>> hold on; bar (binCentres, binCounts, 'r'); hold off
```

4D DATA

Probably the most natural and intuitive example of 4D data is 4DCT which is used for motion management and/or 4D treatment planning in contemporary radiotherapy. 4DCT is a series or sequence of 3DCTs in time. It records time changes of each voxel value throughout a 3D array covering the FOV. A 4DCT data can be expressed and recorded as an array $[x, y, z, t, CTnumber]$ (x,y,z for space, t for time coordinate, and CTnumber for voxel intensity), however, a more natural perspective is a sequence of 3DCTs as mentioned earlier. Typical 4DCT for radiotherapy planning consists of about 10 to 20 3DCTs sampling respiratory cycle so, in terms of a difference to 3D data in previous section, 4D data is about how to organize a series of 3D data. The MATLAB *cell array* data type provides a convenient option. In terms of graphical display, there is no difference in principle to 3D data where a 2D cut only is usually displayed anyway. In case of 4DCT, this is displaying a 2D cut (eg, transverse, coronal, or sagittal) *in time*. Basic data manipulation is analogous to 3D/2D data.

CELL ARRAY

A simple practical example of applying a cell array is an already introduced function to match two 2D-arrays by reference element indices (*match2D*) available in Web Supplement 2.4. The output of this function contains two new generally rectangular 2D-arrays of the same size reflecting the original 2D-arrays overlapping region plus one new reference element indices [*refRow, refCol*] (see Figure 2.8). In general this means three numerical objects but with different dimensionality. Organizing the outcome in a cell array is a simple option for further handling and addressing products of this function. The relevant part of the script is creating a cell array by assigning its three components as soon as available:

```
y = {newA, newB, newRef}
```

Addressing cell array components can be tested using the following example again:

```
>> A = ones (7, 11); refA = [3,6]; B = 2*ones (7, 7); refB = [5, 3];
>> testCell = match2D (A, refA, B, refB);
>> newA = testCell {1}; newB = testCell {2}; newRef = testCell {3};
```

Another example from daily radiotherapy practice is organizing structure contours per the DICOM RT standard where a structure related contour data is recorded as [*x, y*] coordinates of polygonal shaped contour vertexes for each CT slice. Each slice has in general a different number of vertexes. A cell array is a simple option to organize the data when creating a structure 3D mask. This will be demonstrated later in Chapter 3. An extra cell can be then used for recording another structure-related parameter such as a string data type. For a single variable containing, for example, all contoured structure names, strings of variable lengths, a cell array is also a natural option.

STRUCTURE ARRAY

A structure array is an alternative data type to cell array. Both allow storing data of different types and sizes. In this book, a choice between the two is determined by which type is used as an output of a MATLAB inbuilt function such as *dicominfo* extracting header information from DICOM files (for more details see Chapter 3).

An example of a structure array with selected items from a CT transverse slice header information is available in Web Supplement 2.6. To test addressing individual items, the commands in Table 2.13 can be used.

TABLE 2.13 Example of a structure array variable

Command Window	Comments
>> load CTheaderSelect	% load sample structure array variable ...
>> CTheaderSelect	% ... few selected items from a CT image
CTheaderSelect =	DICOM header information. Some items
Format: 'DICOM'	contain string parameters (eg, *Format* or
AcquisitionDate: '20140412'	*ImageComments*), some numerical 1D-arrays
Modality: 'CT'	of variable sizes, etc.
SliceThickness: 1	
ImagePositionPatient: [3 × 1 double]	
ImageOrientationPatient: [6 × 1 double]	
ImageComments: '35s delay @ 3mls'	
Rows: 512	
Columns: 512	
PixelSpacing: [2 × 1 double]	
>> CTheaderSelect.PixelSpacing(1)	% addressing spatial resolution (*PixelSpacing*)
ans =	in row (vertical) dimension in millimeters
0.5059	
>> CTheaderSelect.ImageComments	% addressing *ImageComments* item indicating
ans =	this image was acquired with a delay after a
35s delay @ 3mls	contrast agent administration

TABLE 2.14 Working with a string example

Command Window	Comments
>> sampleString = 'T=90%,PR=87% -> 94%,AR(cm)=-12.92 -> -12.54 4DCT'	% defining a sample string variable using apostrophy signs indicating string start and end (real example)
>> length (sampleString)	% length of string in number of characters
ans =	
49	
>> size (sampleString)	% string as a 1D array of characters
ans =	
1 49	
>> sampleString (length (sampleString))	% addressing last character of the string
ans =	variable
T	
>> percentSigns = find (sampleString == '%')	% finding indices of percent signs within the string variable
percentSigns =	
5 12 19	
>> equalSigns = find (sampleString == '=')	% ... the same for equal signs
equalSigns =	
2 9 27	
>> NominalPhase = str2num (sampleString (equalSigns(1)+1:percentSigns(1)-1))	% extract 4DCT's nominal respiratory phase value and converting it from string to number for further processing
NominalPhase =	
90	

STRING

Besides containing text information, string data type is sometimes used to indicate the value of important parameters. For example, a 4DCT phase bin and amplitude and phase ranges can be recorded in DICOM *SeriesDescription* tag as string. For basic operations such as addressing and replacements, strings are effectively character arrays. Basic operations are illustrated in Table 2.14.

REFERENCES

ICRU Report 83. (2010). Prescribing, recording and reporting intensity-modulated photon-beam therapy (IMRT). *Journal of International Commission on Radiation Units* 10(8).

MATLAB Help R2013a Offline. (2013). MATLAB, The MathWorks, Inc., Massachussets, USA.

CHAPTER 3

Reconstructing Basic DICOM RT Data

LEARNING OUTCOMES

LO 3.1 Explain the purpose and major application areas of DICOM

LO 3.2 Explain the specifics of DICOM RT (radiotherapy module)

LO 3.3 Sort DICOM files by *SeriesInstanceUID* attribute in single-series folders

LO 3.4 Reconstruct a DICOM 3D image series files

LO 3.5 Reconstruct VOIs from a given DICOM RTSTRUCT (structures) file

LO 3.6 Reconstruct 3D dose distributions from a given DICOM RTDOSE (dose) file

LO 3.7 Reconstruct a dose-RT structure image using an example radiotherapy case, calculate DVHs, and compare them with those exported from a TPS

LO 3.8 Explain the importance of verification of reconstruction procedure results: RT structures against reference 3D image, and 3D dose against RT structures (using DVHs)

INTRODUCTION

This chapter contains the essential knowledge to start working with DICOM images and other data specific to radiotherapy (RT). This includes radiation treatment plans, reference images, volumes of interest (VOIs) or simply 'structures', dose distributions, dose-volume histograms, and beam delivery records. Simple algorithms using MATLAB to import and reconstruct DICOM image series including radiotherapy structures and dose are also demonstrated.

DICOM: ALL YOU NEED TO KNOW TO START WORKING

DICOM – Digital Imaging and Communications in Medicine – is *the* international standard for medical images and associated information (ISO 12052:2017). It defines the formats for medical images and associated data and makes the communication of images between devices possible. DICOM is implemented in almost every radiology, cardiology imaging and radiotherapy device (X-ray, CT, magnetic resonance imaging (MRI), ultrasound, etc.), and is increasingly used in devices in other medical domains including ophthalmology and dentistry (NEMA-A).

A simplified view of DICOM images is that they consist of two components: the image itself in a specific image format and the image header. The image header contains relevant clinical and technical information related to the given image. This information, specific to *Service-Object Pair (SOP)*, for example, CT Image Storage, Radiation Therapy Dose Storage, and so on, is organized in individual *attributes*. An attribute example is *'Patient's name'*. Each attribute has a specific numeric code consisting of a pair of 4-digit numbers called a DICOM tag. For example, in the case of a patient's name the tag is *(0010, 0010)*. Attributes are organized in requirement types: *1 or 2 (required), 3 (optional), 1C or 2C (conditional)*. The difference among types is about requirements or conditions related to a given attribute's value length and/or attribute's inclusion based on a given condition (DICOMLookup-A).

RADIOTHERAPY SPECIFIC MODULE: DICOM RT

Radiation therapy is a well defined branch of medicine with a specific workflow and dataflow. DICOM RT modules reflect this by standardizing the storage and transfer of data. Table 3.1 lists current SOPs related to nine data categories that are specific to radiotherapy including various modalities like external photon beam radiotherapy using conventional isocentric linear accelerator, brachytherapy and modern ion radiotherapy.

In general, the contemporary radiotherapy process can be divided into several steps, each is associated with certain data generated and transferred. Figure 3.1 demonstrates key steps of a state-of-the-art external photon beam radiotherapy process. Table 3.2 illustrates the process in a bit more detail with respect to key data generated or transferred as well as associated DICOM modality.

By no means is the external photon beam RT the only RT category. For example, workflow and dataflow for brachytherapy and ion therapy differ from the general process described in Table 3.2. Additionally, even within the external photon beam category there are many alternative modalities to conventional C-arm isocentric clinical linear accelerators and more are in development (eg, MR-guided linacs). Although some features are common to all machines, delivering RT (eg, patient model, RT structures, planned dose distribution), there are some that are specific to each given modality and dose delivery method, which may not be (at least yet) supported by the general DICOM standard (eg, CyberKnife® treatment plan).

TABLE 3.1 DICOM RT module SOPs (DICOMLookup-B)

SOP UID	SOP Name
1.2.840.10008.5.1.4.1.1.481.1	Radiation Therapy Image Storage
1.2.840.10008.5.1.4.1.1.481.2	Radiation Therapy Dose Storage
1.2.840.10008.5.1.4.1.1.481.3	Radiation Therapy Structure Set Storage
1.2.840.10008.5.1.4.1.1.481.4	Radiation Therapy Beams Treatment Record Storage
1.2.840.10008.5.1.4.1.1.481.5	Radiation Therapy Plan Storage
1.2.840.10008.5.1.4.1.1.481.6	Radiation Therapy Brachy Treatment Record Storage
1.2.840.10008.5.1.4.1.1.481.7	Radiation Therapy Treatment Summary Record Storage
1.2.840.10008.5.1.4.1.1.481.8	Radiation Therapy Ion Plan Storage
1.2.840.10008.5.1.4.1.1.481.9	Radiation Therapy Ion Beams Treatment Record Storage

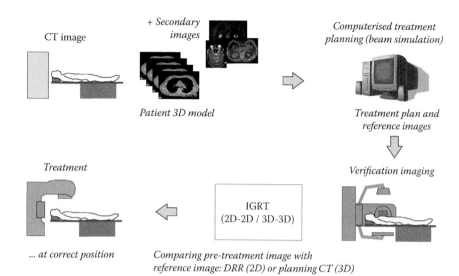

FIGURE 3.1 Contemporary standard 3D conformal image-guided radio therapy (without advanced features such as treatment monitoring, 4D patient model, 4D verification imaging, MR-guided radio-therapy, gating/tracking, real-time plan adaptation, recording and reconstructing delivered, dose, etc.).

TABLE 3.2 Contemporary photon external beam radiotherapy process, related data and DICOM modality (NEMA-B)

RT Process	Related Data	Related DICOM Modality
Diagnose and decision to include RT in treatment	Diagnostic image series including functional exams	CT, MR, PT, XA, …
Imaging for treatment planning (patient model)	Primary CT image series at treatment position on the flat table	CT
	Secondary images to help with target and/or organ definition, motion assessment, etc.	CT, MR, PT, XA, …
Target and critical organs definition (contouring)	Set of 3D contours on patient model defining target and critical structures (volumes)	RTSTRUCT (RT structure set)
Treatment planning	Treatment plan: machine parameters to deliver desired dose distribution	RTPLAN (RT treatment plan)
	Planned dose distribution	RTDOSE (RT dose)
	Beam fluence maps	RTIMAGE
	Reference images (2D, eg, DRR)	RTIMAGE
	Reference images (3D)	CT
Verification imaging (IGRT)	2D kV/MV images (eg, image using EPID or kV imager)	RTIMAGE
	3D kV/MV images (eg, CBCT using kV imager)	CT
Treatment delivery	Recorded dose delivered, IGRT shifts detected/ applied	RTRECORD (RT treatment record)

Notes:
- Modern RT is three-dimensional, ie, based on 3D patient models formed using 3D image series. Hence 3D imaging modalities produce input data.
- Although the primary (model) image is usually CT, there are treatment planning systems using MR instead. Recent developments in MR image-guided radio therapy systems represent probably the most important application.

IMPORTING DICOM DATA

The MATLAB/Image Processing Toolbox contains two key inbuilt functions. They import the:

- DICOM header (metadata): *dicominfo*
- DICOM image (including dose): *dicomread*

In general, one uses *dicomread* in order to obtain array structured data such as 2D or 3D images or 2D or 3D dose field, and *dicominfo* for all other data including, for example, contour points defining a volume structure outline (RT Structure Set) or DVH (RT Plan). The outputs of *dicominfo* are structured arrays with field names that are specific to the corresponding DICOM attribute. See MATLAB Help for details.

Image Series

In a hospital database, each patient with a given name, hospital ID and Date-Of-Birth may have one or more exams using various (imaging) modalities. These exams are referred to as image studies (eg, 'CT abdomen', 'MR head', 'Whole body PET/CT', etc.). In the case of 3D image modalities such as CT, each study consists of one or more image series since there may be more image acquisitions carried out as part of a single study (eg, 'CT abdomen native', 'CT abdomen with contrast', 'CT abdomen at inspiration breath hold', etc.). Localizers or scout images are also stored as a separate image series even if the number of images is just one.

A 3D image series typically consists of a number of transversal, coronal, or sagittal tomograms typically stored as one DICOM file per slice. Normally, the files are transferred to a memory location in the network where they are stored and then await download for various applications or data processing by eligible medical staff. Alternatively, files can be exported to any permitted data storage/transfer medium (CD, DVD, memory stick, etc.). The important practical aspect is that usually DICOM files are not organized in folders based on the study or series. Although some parts of a filename may reflect, for example, slice order, the DICOM file names typically do not represent their contents or acquisition parameters. They may consist of numbers and dots only, or they may contain a specific prefix (eg, 'CT.1.2.246.352.71.3.496497982.502596.20110816120811.dcm'). Quite often file names do not even contain the DCM extension. This means that to organize or reconstruct the data in series or studies their contents must be inspected and organized based on some preference. Table 3.3 illustrates major DICOM attributes allowing the organization of DICOM data in a series to make it available for viewing or processing.

The most relevant attribute used to organize data is the *Series Instance UID*, owing to its mandatory status (type) and series specific uniqueness. Once we organize the DICOM data files in folders based on this attribute, we know that we are working with related data, a single series. For example, a 3D image series is typically stored and exported as a number of files, each containing a single tomogram.

Now we know enough to write a script to organize DICOM files in series specific folders. Let's assume there is a directory containing all image files in DICOM format related

TABLE 3.3   Relevant parameters for sorting DICOM image files to reconstruct image series (DICOMLookup-C)

Tag	Attribute	Type	Description
(0008, 0060)	Modality	1	Type of equipment that originally acquired the data and is used to create the images in this Series
(0020, 000D)	Study Instance UID	1	Unique identifier for the Study
(0020, 0010)	Study ID	2	User or equipment generated Study identifier
(0008, 1030)	Study Description	3	Institution-generated description or classification of the Study (component) performed
(0020, 000E)	Series Instance UID	1	Unique identifier for the Series
(0020, 0011)	Series Number	2	A number that identifies this Series
(0008, 103E)	Series Description	3	User provided description of the Series

to a certain number of image series, and that no related DICOM files are either above or below the file structure level of the given directory. If this were the case, file search must include all relevant directories. However, this is not an over-restrictive prerequisite and to start working with files requires file availability and organization using standard file management features. There may be other, unrelated files in the directory but one must be aware that scanning through more files in a given directory requires more time so it is reasonable to pre-select the data based on information available such as data files type, origin, names, creation time, and so on. This is also the reason why file storage network sites associated with import/export DICOM filters should be maintained. If there is only one application possible, say importing a CT series in a treatment planning system, the data should be removed rather than copied over to ease the job for DICOM import filters or services watching specific data availability (daemons).

Algorithm Scheme

Function Syntax
    sortDCM2series
Function Input/Output

   Prerequisite
      In MATLAB, navigate in the directory containing the DICOM files of interest
   Outcome
      DICOM files of each image series moved in a series–specific folder named by SeriesInstanceUID. The number of folders is identical to the total number of image series through all scanned DICOM files.

   Algorithm Steps

   • Loop: Scanning through all files in current directory
      • If the file is DICOM and contains the *SeriesInstanceUID* attribute and there is no folder named by this *SeriesInstanceUID* then

- – Create a new folder named *SeriesInstanceUID* and move the file in this folder
- If there is the folder where the file belongs, ie, named by given *SeriesInstanceUID*, then
  - – Move the file in this folder
- If the file is not DICOM or does not have *SeriesInstanceUID* attribute then
  - – Do nothing
- Carry on the loop through all files in current directory

The *sortDCM2series* MATLAB script is available in Web Supplement 3.1. Using this function we obtain folders named by respective *SeriesInstanceUID* containing the given series files (images). For better orientation among multiple series it might be helpful to rename folders by a more descriptive series identification. For example, *SeriesNumber*, *ProtocolName*, or *SeriesDescription* are convenient attributes describing key parameters to distinguish image series but note the latter two are type 3, that is, optional attributes and they may or may not be included or be of zero length for a particular modality. An option is to use a combination, for example, 'Series_2' when *ProtocolName* is not available, and, for example, 'Series_2_ t2_tse_tra' when it is. The corresponding algorithm expansion follows: (… algorithm continued)

- Loop: scan through all newly created folders named by *SeriesInstanceUID*
  - Open the first DICOM file in the folder and check if the series specific *SeriesNumber* and *ProtocolName* attributes are present. If they are then …
    - – Rename the folder using one or combination of both parameter values (optionally replace space characters by underscore character to avoid spaces in folder names, or similar)
  - Display the folder/series specific acquisition time contained in the *SeriesTime* DICOM attribute (this may help further with orientation among multiple series)
  - Carry on the loop through all series specific folders

Once we have a single DICOM image series data in one folder we can proceed to data import and reconstruction in MATLAB but before doing so, let's explain a few more essentials. Same as for a DICOM study or series, DICOM file names generally do not represent the rank of the content image within the series so the files need be inspected and sorted based on a relevant parameter. Three relevant DICOM attributes are presented in Table 3.4.

TABLE 3.4 Key parameters related to an image position or orientation relative to patient anatomical reference coordinate system (DICOMLookup-C)

Tag	Attribute	Type	Description
(0020, 1041)	Slice Location	3	Relative position of exposure expressed in mm
(0020, 0037)	Image Orientation Patient	1/1C/2C	The direction cosines of the first row and the first column with respect to the patient
(0020, 0032)	Image Position Patient	1/1C	The x, y and z coordinates of the upper left corner center (of the first voxel transmitted) of the image/frame, in mm

The most simplified view of a 3D image is that it consists of a *series* of 2D images (slices) covering the imaged anatomy. Individual images have their *thickness*, dependent on the image acquisition technique and their *location*. Slice location and thickness are normally equidistant and uniform within the series but there may be *gaps* between slices (see Figure 3.3). And, the slices may or may not be orthogonal with the patient's major anatomical axes. Slice *orientation*, relative to patient anatomy, is recorded in the *ImageOrientationPatient* attribute. It consists of two 3D vectors describing the angle between the first row and the first column and patient reference anatomy. An example of a patient's reference coordinate system XYZ and major anatomical directions (*Anterior, Posterior, Left, Right, Inferior, Superior*) is shown in Figure 3.2. The figure also shows an example of directional cosines (*ImageOrientationPatient* vector) for the image matrix rotated by 45° around patient's longitudinal axis (Z). A normalized (unit length) vector projected under 45° on an orthogonal axis gives $cos\ (45°) = \sqrt{2}/2 = 0.707$. (DicomIsEasy.)

In radiotherapy practice, treatment planning images are usually orthogonal, that is, with the *ImageOrientationPatient* vector consisting of ones and zeros only. Treatment planning systems may not even support importing non-orthogonal images. The reasons for using orthogonal images for the primary image are due to the construction of treatment machines. An isocentric machine with coplanar imaging modality produces an image orthogonal to a patient at standard position. Also registering non-orthogonal secondary image to the primary orthogonal image requires interpolation and some deterioration in quality. So, although oblique images are sometimes better in order to visualize certain anatomical structure, radiotherapy typically works with orthogonal series.

Let's now describe the algorithm for data import and reconstruction in MATLAB. First, let's think about the function output. Of course, we can just leave it to as a 2D or 3D array representing a given image series. However, if one considers the fact that usually we import in MATLAB for some kind of processing, it is reasonable to extract key parameters from the DICOM metadata complementary to the image itself in order to have them at hand for typical operations, such as measuring distance, resize/interpolation, and so on. Selecting parameters depends not only on the application and the user's preference, but also on the

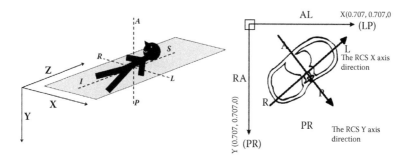

FIGURE 3.2 Patient's reference coordinate system (RCS) vs. major anatomical directions (left). Example of the *ImageOrientationPatient* vector (attribute) for the image rotated by 45° around the patient's longitudinal axis Z (right). (DicomIsEasy.)

patient's position (eg, HFS, Head First Supine), pixel spacing (eg, 0.48×0.48 mm²), slice thickness (eg, 2.5 mm), distance between neighbouring slice positions (eg, 2.5 mm), and the DICOM origin position in XYZ (eg, [–249.7559, –249.7559, 77.5000]) certainly belong to an essential information set complementing the image itself.

Algorithm Scheme

Function Syntax
  importDCMseries
Function Input/Output

  Prerequisite
    In MATLAB navigate in the directory containing files of a single DICOM series
  Output
    dcmImgImport a MATLAB cell array with the following elements:
      1. *imgMatrix* (image 2D or 3D array)
      2. *Modality* (string: CT, PT, MR, XA, …)
      3. *PatientPosition* (string: HFS, FFS, HFP, FFP …)
      4. *SliceOrientation* (string: transversal, coronal, sagittal)
      5. *ImageOrientationPatient* (6-element vector)
      6. *PixelSpacing* (2-element vector for direction of rows and columns)
      7. *SliceThickness* (one value if uniform within the series, minimum and maximum otherwise)
      8. *SliceSpacing* (distance between neighbouring slice location, one value if uniform within the series, minimum and maximum otherwise)
      9. *OriginXYZ* (3-element vector indicating DICOM origin reference point)

Algorithm Steps

- Loop: Scan through all files in current directory
  - For each DICOM file record the file rank (in the directory) and corresponding *ImagePositionPatient* attribute in the [$n \times 4$] file-index vs. slice-location array (*SliceLocationFile*)
      Note: *SliceLocation* attribute may not be available, for example, for CBCT from a therapy machine kV imaging system
  - Also record number of rows and columns (*Rows* and *Columns* attributes) for the consistency check
- Carry on the loop through all files in current directory
- If number of rows or columns is inconsistent then generate error and terminate the program
- From the first DICOM file in the directory, ie, the first *n*-index of the array, extract the number of (image) rows and columns, from the *Rows* and *Columns* attribute,

Modality, ImageOrientationPatient and *PatientPosition*, using respective attributes. The number of slices is equal to the number of the single series DICOM files in the directory. Note these are attributes whose values should be the same for all image files within the series

- Pre-allocate the image matrix: *zeros (Rows, Columns, Slices)*
- Compare *round (abs (ImageOrientationPatient))* with vector values corresponding to slices orientation transversal, coronal or sagittal (see Table 3.5)

Rounding gets the vector to the closest orthogonal configuration to record as one of the output parameters in the sense of the closest match. Accurate information about slice orientation gives only the full *ImageOrientationPatient*.

- KEY OPERATION! Sort the $[n \times 4]$ file-index vs. *ImagePositionPatient* array based on the *ImagePositionPatient* attribute

Note that if the *SliceLocation* attribute indicating relative slice position within the series is present, it can be used for slice file sorting in order to obtain the imported image series with the correct slice order. In addition, for the orthogonal image series (see Figure 3.3) the *SliceLocation* value equals the corresponding *ImagePositionPatient* vector coordinate, for example, the third, Z-coordinate, for transversal series. So, in principle it is possible to use either *SliceLocation* or the appropriate *ImagePositionPatient* coordinate for slice sorting in most situations. However, there are situations when no *SliceLocation* attribute is present (eg, radiotherapy CBCT) or images are not 100% orthogonal (typical for MR diagnostic series), that is, *ImageOrientationPatient* vector(s) do not consist of zeros and ones exclusively. So to cover both possible options, that is, non-orthogonal images and no *SliceLocation* attribute available, one can use a calculated relative 1D coordinate of a given image along the straight line connecting all *ImagePositionPatient* points. From analytical geometry a straight line is determined by the parametric equation

$$X_i = X_1 + k \times \frac{X_2 - X_1}{norm(X_2 - X_1)}$$

TABLE 3.5   Examples of an *ImageOrientationPatient* vector for common orthogonal slice orientations supported by a commercial treatment planning system (CyberKnife PEG)

Slice Orientation	Image Orientation Patient
Transversal (Z)	(1, 0, 0) (0, 1, 0)
Transversal (Z)	(−1, 0, 0) (0, 1, 0)
Transversal (Z)	(1, 0, 0) (0, −1, 0)
Transversal (Z)	(−1, 0, 0) (0, −1, 0)
Coronal (Y)	(1, 0, 0) (0, 0, −1)
Sagittal (X)	(0, 1, 0) (0, 0, −1)

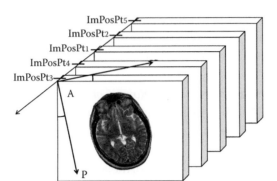

FIGURE 3.3   The slices come in DICOM files whose names do not necessarily reflect their position within the series so sorting based on *SliceLocation* or *ImagePositionPatient* attributes is required after the import. Also note the difference between slice *Spacing* and slice *Thickness*.

where X_i is a general point on the straight line, X_1 is an arbitrary origin, $X_2 - X_1$ is an arbitrary directional vector pointing from X_1 to X_2. With this vector normalized by its length (*norm*), the k parameter gives a 1D coordinate of a given X_i point along the straight line originating in X_1 in direction of X_2. In this example it does not matter which X_1 and X_2 points (*ImagingPositionPatient*) will be used, so picking up the first two DICOM files is an efficient choice. Expand the array by one column indicating the k parameter and sort images using the *sortrows* command. Note now we have the $[n \times 5]$ array of file directory indices with ascending slice-location coordinate.

- Loop: Scanning through all files in a current directory in the order given by the slice-location coordinate
  - For each DICOM file record the *SliceThickness* and *PixelSpacing* attribute values in an appropriate 1D array (*SliceThicknessSeries, PixelSpacingSeries*)
  - Using the *dicomread* import a given image plane and, when available, rescale adequately using the *RescaleSlope* and *RescaleIntercept* attributes
    Note: *RescaleSlope* and *RescaleInterest* convert stored values (SV) into output units (OU) using *OU = RescaleSlope * SV + RescaleIntercept*. Example of OU are Hounsfield units for CT (DICOMLookup-C)
    Carry on the loop …

Note now we have imported the 3D image matrix

- Check selected parameters for consistency within the image series
  - Check consistency for selected parameters: *PixelSpacing* and *SliceThickness* by inspecting the respective arrays with the recorded values of all images in the series
    - Both parameters are expected constant within the series so the given array differential vector is expected to be zero. However, due to rounding some

minimal variation it is possible hence the condition similar to the following one is reasonable:

```
max (abs (diff (SliceThicknessSeries))) > 0.0001
```

- – If the condition is met, the parameter values are inconsistent, then
  - – Display warning
  - – Record minimum and maximum value
- – Otherwise
  - – Record single uniform parameter value
- • Check consistency for the *SliceSpacing* parameter
  - – The difference in location between any two neighbours in *SliceLocationSeries* is expected to be constant in radiotherapy. This means that twice the differential should be zero again but a small value can be used instead for similar reasons as above:

```
max (abs (diff (diff (SliceLocation)))) > 0.0001
```

- – Action analogous to the two parameters above
- • Assign relevant parameters to the output cell array:
  dcmImgImport {i} = a given parameter value (*imgMatrix, Modality, PatientPosition, SliceOrientation, ImageOrientationPatient, PixelSpacing, SliceThickness, SliceSpacing, OriginXYZ*)

There are a few added verification steps including examples of generating warning or error messages for screen display. The *importDCMseries* function script is available in Web Supplement 3.1. To test both functions one can use the following command sequence. The *CTexamples* folder in Web Supplement 3.1 contains a mixture of two CT image series (1 diagnostic CT of a brain, 1 CBCT from a treatment machine), so the task is to sort all files in two folders and import both series in MATLAB.

Make sure MATLAB knows where to find both function scripts using the *addpath* command and navigate in the folder containing the mixture of DICOM files. Then …

```
>> sortDCM2series
```

… navigate in the first newly created folder and run

```
>> CT1import = importDCMseries;
```

… and alternatively for the second folder and image series.

To verify correct slice ordering, view any plane orthogonal to the imaging plane. For example, for the *CT1* use the sagittal plane through the central column (of 512):

```
>> figure; CT1 = CT1import {1}; imagesc (squeeze (CT1_3D (:,
   256,:))); colormap gray
```

… and alternatively for the *CT2*.

Secondary Image Series (MRI, PET, ...)

The secondary image series became standard in modern radiotherapy. MR image series are commonly used for, for example, prostate and brain treatment sites and PET/CT is considered standard for lung treatment. Purposes for the use of a secondary image series of modalities other than X-ray CT for treatment planning are often:

- Improved target definition for better contrast (eg, MR) and/or functional imaging (eg, fMRI, PET/CT, XSA, etc.)

- Improved OAR definition for the same as above (eg, nerves and MR)

However, secondary images for treatment planning do not mean necessarily other than CT modality. Also CT images can serve well as secondary images for treatment planning:

- CT with contrast agent to improve target definition again

- 4DCT for motion management (intra-fractional motion)

- Other available CT series to assess inter-fractional variation, and so on

Although the most common image acquisition is transverse, this is not always the case. In MR, for example, sagittal as well as coronal planes are commonly acquired together within one study. Both CT and MR sample image series, both transverse and non-transverse, are available in Web Supplement 3.1 to test the *importDCMseries* function also for other than CT modality and non-transverse acquisitions.

Volumes of Interest (RT Structures)

Together with the (primary) image series representing the patient's model, the radiotherapy structure set forms the base for modern computerized 3D radiotherapy treatment planning. The patient 3D model, still almost exclusively based on X-ray CT today, serves the two main purposes:

- Provide the model platform to define target as well as critical volumes to focus and to avoid treatment dose, respectively

- Provide a 3D map of the distribution of electron density used by dose calculation algorithms to calculate the distribution of dose. Some modern algorithms (eg, based on Monte Carlo or particle transport modeling) require also distribution of material composition. In such cases a CT image weighted 'just' by electron density has to be accompanied by material tables. For example, dual energy CT helps in distinguishing between materials with similar electron density

For the purposes of treatment planning, that is, simulation and optimization of radiation beams towards clinically acceptable dose distribution, the algorithms need to know which voxels within the patient's model form a target or a critical structure. This is also essential for the inverse planning approach typical for modern fluence modulated techniques such as IMRT or VMAT.

Leaving auto-segmentation algorithms aside, the structures or *volumes of interest* (VOIs) are still presently defined by contouring. Since standard three-dimensional image viewing is via two-dimensional display of slices, the contouring process also consists of defining 2D contours on each slice to include a given structure. A set of individual points that form the contour is given to the computer by registering the position of the mouse, electronic pen or a similar tool. All contour points for a given slice and structure generally have identical slice-location coordinate. So, the organized set of the *XYZ* coordinates of contour points is the form the DICOM standard stores the structure set. The *XYZ* coordinate system for the contour set is identical with the coordinate system of the underlying image.

There are few contour types specified in the *ContourGeometryType* attribute: *point*, *open_planar*, *open_nonplanar* and *closed_planar*. Most radiotherapy structure contours (volumes) are formed by series of *closed_planar* contours (NEMA-C).

Although registered secondary images (MR, PET/CT, …) are often used to help visualize a particular structure or specific physiological process, a given contour set is linked exclusively to the primary image. This exclusive association is ensured by cross-referencing the source image (CT series) identifier in the structure set file, namely the *SeriesInstanceUID* attribute included to *ReferencedFrameOfReferenceSequence*.

For reasons mentioned earlier, the patient's model is based on an orthogonal (CT) image series, typically transversal. Supporting secondary images may be sagittal or coronal in order to better visualize a particular structure of interest (typical for MR). Therefore, some modern treatment planning systems enable contouring on sagittal or coronal slices too. Taking this into consideration when writing scripts for reconstructing structures in MATLAB yields a more robust solution.

Let's proceed in formulating the example algorithms for importing and reconstructing RT structures in MATLAB. As always, let's think about the inputs and outputs first. The first input is obvious, the DICOM RT structure set file. Regarding the output, one needs to consider the purpose which is the structures reconstruction on the reference image series. Therefore, an obvious outcome should be a 3D binary array mask (*mask3D*) of the same size as the reference (3D) image (*refImage*) with *ones* indicating voxels belonging to a given structure and *zeros* otherwise. However, such form of output is demanding on computer memory (eg, a 3D matrix of $512 \times 512 \times 100$), so it is practical to store the full information in the form of, for example, linear indices of ones (let us call this *mask1D*) taking just a fraction of computer memory, based on a given structure volume. Knowing the size of the reference image matrix it is then trivial to reconstruct the full 3D mask using:

```
>> mask3D = zeros (size (refImage)); mask3D (mask1D) = 1;
```

In order to convert the *XYZ* contour triplets to the Row-Column-Slice (*RCS*) – matrix voxel triplets, the reference *XYZ* coordinate system is required. The *ImagePositionPatient* attribute of the reference image first slice (with the lowest slice coordinate) provides the coordinate system origin. Reference image pixel size (*PixelSize*) and slice spacing (*SliceSpacing*) are the next parameters required for the conversion.

$$
\begin{bmatrix} c \\ r \\ s \end{bmatrix} = round \left(\begin{bmatrix} (x - x_0)/PixelSize_x \\ (y - y_0)/PixelSize_y \\ (z - z_0)/SliceSpacing \end{bmatrix} \right)
$$

This conversion applies for the patient orientation HFS and a CT matrix obtained using the *importDCMseries* described earlier. See Table 6.2 for more patient orientations and respective relation to matrix coordinates of the reconstructed 3D image that have to be considered for alternative or general conversion.

In general, a given structure set (contained in one DICOM file) contains contour data for all defined structures. So, in the interest of efficacy, the function should reconstruct only one selected structure identified by its structure ID within the set specified on the input. If more structures need reconstruction, then this can be organized via an appropriate external script.

Considering the information so far, the overall framework for importing and reconstructing structures in MATLAB including initial verification can be split into the following phases:

- Navigate in folder with one RT structure set and all reference image DICOM files

- List the structure set sequence contained in the structure file on the computer screen (structure ID, structure name)

- For selected structure (ID) and the RT structure, set filename and call the single-structure reconstruction function in order to obtain a binary mask in the form of linear indices of voxels (*mask1D*) forming the given structure on the reference image (patient model)

- From the linear indices reconstruct the 3D binary mask array (*mask3D*)

- Consider comparing the mask with the reference image using MATLAB 2D plotting features. As the body contour ('skin') is usually included in the structure set, it is the most natural choice to check the reconstructed mask shape and orientation

Algorithm Scheme (Listing Structure IDs and Names as Stored in the DICOM File)

Function Syntax
    RSdicom2names (RSfilename)
Function Input/Output

Input

    RSfilename   DICOM RT structures filename (includes the structure name, ID within the contour set and contour data)

Outcome   List of the structure set sequence (structure ID, structure name) on the computer screen

Algorithm Steps

- Import the RS file in MATLAB using *dicominfo*

```
RS = dicominfo(RSfilename);
```

- Obtain number of structures (VOIs) contained

```
length (fieldnames (RS.ROIContourSequence))
```

- Loop: for each structure index (*i*) extract its *ROInumber* and *ROIname* attributes

```
eval (['RS.StructureSetROISequence.Item_',num2str(i),'.ROINumber']);
eval (['RS.StructureSetROISequence.Item_',num2str(i),'.ROIName']);
```

- …and display the *ROInumber* vs. *ROIname* pair on-screen, eg,

```
ID = 1    Structure name = BODY_P1
ID = 2    Structure name = CouchSurface_P1
…etc.
```

Now, having a structure of interest ID, we can proceed to the given contour set reconstruction. A general scheme is illustrated in Figure 3.4.

Algorithm Scheme (Reconstructing a Given Structure in the Form of Voxel Linear Indices)

Function Syntax
    RSdicom2mask (RSfilename, ROInumber)
Function Input/Output

    Input

        RSfilename   DICOM RT structures filename (includes the structure name, ID within the contour set and contour data)

        ROInumber   A given structure ID in the structure set sequence

    Output

        mask1D   A 1D array of linear indices of voxels belonging to a given structure within the 3D reference image matrix

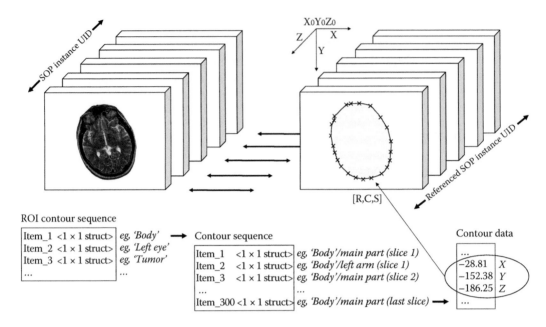

FIGURE 3.4 RT structures reconstruction scheme: each contour point *XYZ* from a given *ContourSequence* of a given structure (*ROIContourSequence*) is reconstructed as the corresponding *RCS* voxel in the reference image 3D array. *ContourSequence*-specific *ReferencedSOPInstanceUID* helps matching respective slice.

Prerequisite

Organize the reference (CT) image series and the DICOM RTSTRUCT files in one folder to ease files handling in the algorithm description below

Algorithm Steps

Note: In order to distinguish arbitrary variable names from the DICOM attributes, either VAR or ATT are indicated in brackets behind each parameter within the algorithm description

- Scan through the folder with DICOM (CT) reference image series files and for each file record
  - The file rank (index) in the folder
  - The file (slice) specific *ImageOriginPatient* vector
  - The file (slice) specific *SOPInstanceUID*

and calculate

  - The slice location the same way as described previously for the *importDCMseries* function

The data can be recorded in a cell array (*SliceLocationFile* in the script)

- Sort the files (slices) records by the slice location parameter
  - From any file (slice) record of *ImagePositionPatient* define the first and second coordinate *CTx0, CTy0*. This represents the *XY* coordinate system origin for 2D conversions within each slice. Within the series these coordinates are identical for all the slices

- from the respective DICOM attribute extract the number of rows (*CTrows*), columns (*CTcolumns*), number of slices *NoOfCTs = length (SliceLocationFile)* and *PixelSize* (*PixelSpacing(1)* attribute)
- import the DICOM RTSTRUCT file

```
RSinfo = dicominfo (RSfilename);
```

- Loop for each *ContourGroupIndex* (VAR), ie, the *Item_X* (ATT) in the *ContourSequence* (ATT) and record *ContourGroupRefInstance* (VAR), that is, the *ReferencedSOPInstanceUID* (ATT)
  - Search through the *SliceLocationFile* (VAR) for the *ContourGroupRefInstance* and record corresponding *SliceRank* (VAR) within the sorted reference 3D image matrix
  - Extract the *XYZ ContourGroupDATA* (VAR) from the respective *ContourData* (ATT)
  - Convert the *XYZ* in *RCS* and record it in the *ContourGroupDATA_Voxels* (VAR)

```
ContourGroupDATA_Voxels(CountourPointNo, 1) = round
((ContourGroupDATA (3*CountourPointNo-2, 1) - CTx0)/PixelSize);
ContourGroupDATA_Voxels(CountourPointNo, 2) = round
((ContourGroupDATA (3*CountourPointNo-1, 1) - CTy0)/PixelSize);
ContourGroupDATA_Voxels (CountourPointNo, 3) = SliceRank;
```

- Store the contour specific voxel *RCS* coordinates in the cell array

```
ContourGroupDATA_Structure {ContourGroupIndex} =
ContourGroupDATA_Voxels;
```

- Record the *ContourGroupIndex* vs. the *SliceRank* for later quick link between the contour voxels and the given reference slice …

```
ContourGroupXCTslice (ContourGroupIndex,:) = [ContourGroupIndex,
ContourGroupDATA_Voxels (1,3)];
```

- Loop for each reference image slice *CTsliceIndex* (VAR)
  - From the *ContourGroupXCTslice* (VAR) find the associated *ContourGroupDATA_Structure* (VAR) indices …
- Loop for each contour group on current slice
  - Create a zero 2D mask [*CTrows* x *CTcolumns*]
  - From the relevant contour group data reconstruct the current slice-specific 2D mask using

```
A = zeros(CTrows,CTcolumns);     contourMask = roipoly
(A,ContourGroupDATA_CT(:,1),ContourGroupDATA_CT(:,2));
```

  – Sum up all current slice allocated 2D masks to obtain the integral (multi-contour group) 2D mask for the current slice
  – Consecutively add up to the final *mask1D* (VAR) vector linear voxel indices of all non-zero voxels of the current slice-specific 2D mask using also the reference (CT) matrix dimensions and current slice index in it

There can be gaps between reference image slices with a contour defined, but this is uncommon since the structures to deal with in MATLAB come from a TPS export, that is, structures included are supposed to be processed. Processing includes contour interpolation on slices where the contour is not defined explicitly. So, when reconstructing a given structure 3D mask, it is usually unnecessary to fill the potential gaps between slices using interpolation. It is simple to check for such gaps (individual slice contour point sequence is or is not empty) but in order to decide whether this is intentional or interpolation is needed requires user's decision. For structures with multiple contour sets, that is, multiple sub-volumes, they may require merging after the import depending on how a particular TPS exports. However, this should be unnecessary since typically all contour sets should be exported under one associated structure (*ROIContourSequence*).

Having the reference image 3D matrix and reconstructed binary 3D mask for a given structure, it is simple to verify the correctness of the reconstructing process by displaying a selected mask 2D cut through the mask 3D matrix as a 1-contour plot on the background of the equivalent 2D cut through the reference image matrix. Of course, this requires using structures clearly identifiable in the reference CT image, for example, body (skin), eyes, kidneys, and so on. An example using the sample data is available in Web Supplement 3.4 and is presented in 'Reconstructing 3D Dose Distribution' below.

The *RSdicom2names* (displaying stored RT structures names) and *RSdicom2mask* (reconstructing stored RT structures binary 1D masks) MATLAB script examples together with example clinical data are available in Web Supplement 3.2.

Dose

Now, having reconstructed a reference image series and RT structures, we can now import and reconstruct the next radiotherapy specific modality – the dose. Unlike the reference image series and the structures, the range of RT dose data stored in the DICOM file is usually defined by the user during the treatment planning process or export. Depending on the treatment planning system, the dose calculation grid and spatial resolution can be defined entirely independent of the reference image resolution, slice thickness or spacing, the outer body (skin) contour, and so on. That is why, normally, one has to process the dose data to reconstruct and match it to the reference image set and associated RT structures to reproduce standard treatment planning system conditions.

There are many model situations when exporting the dose from a treatment planning system is required. For example,

- To obtain the reference dose plane from a treatment plan on a phantom for comparison with a 2D dose measurement (standard situation for current approach to IMRT plan verification)

- To obtain the reference 3D dose on a phantom for the same purpose as above

- To obtain the reference 3D dose on a patient model for plan verification based on reconstructed 3D dose from 2D (transit) measurements

- To obtain the reference 3D dose on a patient model for alternative dose calculation as part of plan QA

- To obtain the 3D dose on a patient model for treatment plan sum assessments, statistics and research, and so on

The dose distribution imports from the DICOM RTDOSE file as a numeric array by using the *dicomread* inbuilt function. All additional parameters stored in the file are imported using the *dicominfo* function. Let's assume importing from a given DICOM RTDOSE file using the named functions:

```
>> dose = dicomread ('RDfilename.dcm');
>> doseInfo = dicominfo ('RDfilename.dcm');
```

Then for relevant information content and for further processing the attribute examples listed in Table 3.6 may be essential.

In order to obtain the imported dose array in appropriate dose units, the array must be rescaled using the *DoseGridScaling* factor. Following the example commands above then:

```
>> dose = double (dose) * infoDose.DoseGridScaling
```

TABLE 3.6 Essential parameters required to reconstruct RT dose planes or volumes in MATLAB

DICOM Attribute	Comment
doseInfo.DoseUnits	Exported dose units, eg, 'GY'
doseInfo.DoseGridScaling	Dose scaling factor to multiply the imported dose array to convert values in appropriate dose units
doseInfo.PixelSpacing	Dose plane (2D) or dose frame (3D) pixel spacing in millimeters, eg, [2.5, 2.5]
doseInfo.Rows/Columns	Number of rows/columns of a dose plane (2D) or a dose frame (3D)
doseInfo.ImagePositionPatient	Exported dose grid reference *XYZ* coordinates
doseInfo.ImageOrientationPatient	Exported dose grid orientation vector
doseInfo.NumberOfFrames	Exported dose number of frames (slices) – for 3D dose
doseInfo.GridFrameOffsetVector	Exported dose frames (slices) offset from the origin vector (slice spacing) for 3D dose
doseInfo.DVHSequence	Exported dose volume histograms …

The *dicomread* imports 3D dose as a 4D-array [*rows, columns, 1, frames/slices*]. Conversion into a convenient 3D matrix is trivial by removing the singleton dimension (dose):

```
>> dose = squeeze (dose);
```

Depending on the treatment planning system or export parameters, the dose can be the total, that is, the sum contribution of all beams, or a specific single field.

The major field of application for the utilization of native RT dose matrix is a direct comparison with a measured dose. The next unique data contained in the DICOM RTDOSE file, dose volume histograms, provide a wide application range including, for example, TCP/NTCP modeling, and so on. Other applications typically require dose registration with a reference image (CT) and RT structures (see the section 'Reconstructing 3D Dose Distribution' later in this chapter).

DVH

For a general introduction to dose volume histograms and related basic operations, please refer to Chapter 2. DVHs of structures selected and generated during export from a TPS are stored in the DICOM RTDOSE file under the DVHSequence attribute. In the next section we will demonstrate an algorithm to extract DVH data from a given RTDOSE file. Let us create a MATLAB function with a DICOM RTDOSE file on its input and a numeric 2D table on the output. The first column of the table contains dose (bins). The following columns include corresponding structure specific partial volumes. Essential related DICOM attributes are listed in Table 3.7.

TABLE 3.7 The key parameters required to reconstruct DVHs in MATLAB (NEMA-D)

DICOM Attribute	Comment
DVHType	Type of DVH. Enumerated Values: DIFFERENTIAL (differential dose-volume histogram), CUMULATIVE (cumulative dose-volume histogram), NATURAL (natural dose volume histogram)
DoseUnits	Dose axis units. Enumerated Values: GY, RELATIVE (dose relative to reference value specified in DVH Normalization Dose Value)
DVHVolumeUnits	Volume axis units. Defined Terms: CM3, PERCENT
DVHSequence	Sequence of DVHs stored
DVHNumberOfBins	Number of bins n used for DVH Data storage
DVHData	A data stream describing the dose bin widths D_n and associated volumes V_n in DVH Volume Units in the order $D_1 V_1, D_2 V_2, \dots D_n V_n$
DVHDoseScaling	Scaling factor that when multiplied by the dose bin widths found in DVH Data, yields dose bin widths in the dose units as specified by Dose Units

Algorithm Scheme (Extracting DVH Data)

Function Syntax
  RDdicom2dvh (RDfilename, ROInumber)
Function Input/Output
  Input
    RDfilename   DICOM RTDOSE filename
    ROInumber   A given structure ID in the structure set sequence
  Output
    DVHdata   A 2D array of dose in the first column and corresponding structure-specific partial volume in the second column. Number of rows equals number of dose bins

Algorithm Steps

- Import the RTDOSE file in MATLAB using *dicominfo*. Let's name the structure array variable *RDinfo*.
- Get all stored structures (*i*) their *ReferencedROINumber* using

```
eval(['RDinfo.DVHSequence.Item_',num2str(i),'.
DVHReferencedROISequence.Item_1.ReferencedROINumber'])
```

  and find the index (*I*) corresponding to the structure's *ROInumber* on the input to reconstruct
- Get the *DVHData* and sort it in required 2-column format

```
DVHData = eval (['RDinfo.DVHSequence.Item_',num2str(I),'.
DVHData']);
```

  - Extract the dose bin widths, do the cumulative sum and multiply by *DVHDoseScaling* factor to obtain absolute dose bins vector

```
DVHDoseBins = cumsum (DVHData (1:2:end)) *
eval(['RDinfo.DVHSequence.Item_',num2str(I),'.DVHDoseScaling]);
```

  - Extract the volumes: *DVHVolumes = DVHData (2:2:end)*;
  - Organize the data: *DVH(:, 1) = DVHDoseBins; DVH(:, 2) = DVHVolumes*;
- If the *DoseUnits* is RELATIVE then the dose bins can be converted in GY by multiplication by the *DVHDoseNormalizationValue*
- If the *DVHVolumeUnit* is PERCENT, then the volume values can be converted in *cm3* using the *ROIVolume* stored in the DICOM RTSTRUCT. This means that another DICOM file is required so, considering the decision about the inputs and outputs of the DVH import function, this has to be dealt with externally

The *RDdicom2dose* (reconstruction dose) and *RDdicom2dvh* (reconstructing DVH) MATLAB script examples, together with example clinical data, are available in Web Supplement 3.3.

Other DICOM RT Data in Radiotherapy

From all other DICOM RT modalities listed in Table 3.1, let's present the one which is useful in a standard conventional radiotherapy department: DICOM RTRECORD. Every RT treatment machine operated in clinical mode records delivered parameters. This is not necessarily the dose as independent measurements of this ultimate parameter cannot be considered standard yet, instead there are many other treatment parameters such as beam monitor units (MU), treatment time, and so on. DICOM RTRECORD represents a standard for recording parameters from treatment on a RT machine. There are practical applications of having a RT machine supporting the export of treatment parameters using this DICOM modality. For example, if a RT department operates multiple treatment machines from various manufacturers and only one, for example, Record & Verify (R&V) or Oncology Information System (OIS), using the DICOM RTRECORD standard, it is possible to schedule, verify and update a given patient's treatment. This is possible even when a given machine is otherwise very different from the machines a given R&V system is primarily designed for. Of course, this is only possible when there is a supported interface between both parties, but the DICOM standard enables this in principle. Due to possible large differences among RT machines in terms of treatment delivery parameters (some use MUs, some use treatment time, some are isocentric, some not, beam limiting devices vary a lot, etc.) only some can be used for patient record updates. Even within a single vendor equipment the DICOM RTRECORD might help in situations when the primary database server is not available for any reason. So having an option to handle treatment delivery using files following the DICOM RTPLAN (to deliver) and DICOM RTRECORD (to record) standard, may be useful. Of course, there are many more possible applications of treatment records organized in a standard file downloadable from a given machine control computer(s) – for example, research, delivered dose reconstruction, and so on. In the following paragraphs we will present some essential DICOM attributes specific to the RTRECORD modality.

Similar to other modalities, a DICOM RTRECORD file can be imported in MATLAB easily by a dedicated *dicominfo* inbuilt function. A sample file (*RTRECORDexample.dcm*) is available in Web Supplement 3.4.

```
>> RTinfo = dicominfo ('RTRECORDexample.dcm')
```

The minimum information content for a useful application in the patient's treatment record update that one can consider includes:

- Patient identification
- Delivered treatment plan identification
- Delivered fraction number

- Delivered fraction dose

- Delivered (beam specific) MU/time

Relevant key DICOM attributes named by MATLAB as per the current DICOM dictionary are indicated below. Type

```
>> help dicomdict
```

in MATLAB command window for more info about DICOM dictionary.

```
>> RTinfo.PatientName %
```
(alternatively for *PatientID*, *TreatmentDate*, *TreatmentTime*, *NumberOfFractionsPlanned*, etc.)

provides essential information about treatment delivery.

The other attributes are more technical and may play an important role in patient safety in terms of correct and accurate update of treatment records:

- *ReferencedRTPlanSequence* containing the *ReferencedSOPInstanceUID* relating the given RTRECORD to corresponding RT treatment plan

- *TreatmentSessionBeamSequence*

Information-rich attribute with many treatment parameters including, for example, treatment couch position, gantry rotation angles, and so on.

- *CurrentFractionNumber*

    … crucial parameter for treatment records update

- *ReferencedCalculatedDoseReferenceSequence*

    … a given fraction dose as per associated treatment plan. Not the actually measured dose

- *ReferencedVerificationImageSequence*

    … references to verification images (IGRT) applied during the given treatment

- *SpecifiedPrimaryMeterset (SpecifiedSecondaryMeterset, SpecifiedTreatmentTime)*

    … expected beam MUs or time

- *DeliveredPrimaryMeterset (DeliveredSecondaryMeterset, DeliveredTreatment Time)*

    … measured beam MUs or time. Based on the difference between measured and expected beam dose quantifier meterset the given fraction can acquire status as 'undelivered', 'delivered' or 'partially delivered', or similar.

- *ControlPointDeliverySequence*

  … details of the machine parameters in time – also for dynamic treatments. eg, *GantryAngle, BeamLimitingDevicePositionSequence*, etc.

RECONSTRUCTING 3D DOSE DISTRIBUTION

Now we have scripts for importing DICOM image series (*importDCMseries*), displaying stored RT structures names (*RSdicom2names*), producing selected structures binary masks (*RSdicom2mask*), importing DICOM dose distribution (*RDdicom2dose*) and constructing DVH (*RDdicom2dvh*) so it is possible to test them using example data.

The problem where we can demonstrate the basic application of created scripts is reconstructing 3D image, dose and RT structures. Correctness will be assessed by comparing DVHs, also demonstrating high sensitivity of DVH to geometric accuracy.

The problem can be split in the following steps:

- Prerequisite: Let's have a TPS with DICOM RT export capabilities, import CT series, define some structures (VOIs) and create some RT plan to overlay the dose distribution on the patient (CT) model. Then perform DICOM export of images, plan, RT structures and RT dose (including DVHs) and place all files in one data folder. Example data is available in Web Supplement 3.5.

Then …

- Import the reference (CT) image series in MATLAB

- Display the list of structures exported

- Reconstruct linear structure masks of selected structure(s) in MATLAB

- Import 3D dose in MATLAB

- Register 3D dose with the reference (CT) image

- Calculate DVH(s) of selected structure(s)

- Import reference (TPS-calculated) DVH(s) of selected structures

- Compare calculated and reference DVH(s) to verify correct image-structure mask-dose registration

There are two operations uncovered by scripts or algorithms presented in this chapter: *registering 3D dose with CT* followed by *own calculation of DVHs*. The operations will be presented in the next. Reconstructing the image-RT structure dose from DICOM exports forms the basis for more advanced applications such as multi-dimensional comparison to measurement data, treatment plan sums using rigid image registration, 3D dose calculation, and so on. A simple plan sum and simple 3D dose calculation are demonstrated in Chapters 5 and 7, respectively.

Let's have a DICOM CT image series files and one related DICOM RT structures file in one folder. Then import the CT image series using

```
>> CTimport = importDCMseries;
```

Extract the imported reference image 3D array (see the section 'Importing DICOM Data/Image Series' earlier in this chapter).

```
>> RefImage = CTimport {1};
```

List the names of structures stored in the DICOM RTSTRUCT file:

```
>> RSdicom2names ('RSfileSample_1.dcm')
   'ID = 1    Structure name = BODY_P1'
   'ID = 2    Structure name = CouchSurface_P1'
   'ID = 3    Structure name = CouchInterior_P1'
```

There is only the outer body and the treatment couch stored in the file. Use the structure 1, 'body', to test reconstructing the structure. Let's obtain a linear voxel index sequence first:

```
>> maskBody1D = RSdicom2mask ('RSfileSample.dcm', 1);
```

From the linear indices reconstruct the 3D binary mask array (*mask3D*)

```
>> maskBody3D = zeros (size(RefImage)); maskBody3D (maskBody1D) = 1;
```

And finally, compare the 3D mask binary array with the reference image using the coronal cut display:

```
>> figure (1); imagesc (squeeze (RefImage (floor (size (RefImage,
1)/2),:,:))); colormap gray
>> hold on; contour (squeeze (maskBody3D (floor (size (RefImage,
1)/2),:,:)), 1, 'g');
```

This displays the coronal plane in the mid vertical range of the *RefImage* in grayscale and the same plane of the *maskBody3D* as one green contour, so it is easy to verify correctness of the data reconstruction.

```
>> figure (2); imagesc (squeeze (maskBody3D (floor (size
(RefImage, 1)/2),:,:))); colormap gray
```

… and similar for alternative anatomical planes.

Note that the 3D mask as presented in *figure (2)* represents a higher form of structures reconstruction since it does not only reproduce contours as defined and stored in the

DICOM RTSTRUCT file, but it adds filling the assumed (standard) CLOSED_PLANAR contours to address all voxels of a given volume. *figure (1)* demonstrates visual verification of the reconstructed structure using the 1-contour plot explained previously.

Matching the Dose with the CT Model

Unlike RT structure contour points, which import as XYZ triplets, the RT dose imports as an array of dose values covering a calculation grid set in a given TPS where dose was calculated before export. As calculation grid location, size and resolution are typically defined by the user, the data resampling and registration using a reference point is required to match the dose with the reference image matrix. The key DICOM attributes relevant to this task are listed in Table 3.6 (see the section 'Importing DICOM Data/Dose' earlier in this chapter).

Assuming we already have imported the reference (CT) image series using own *importD-CMseries* function (see the *CTimport* variable in the section 'Importing DICOM Data/Image Series' earlier in this chapter). The next key parameters required to register the dose are:

```
CTPixelSpacing = CTimport {6}; CTSliceSpacing = CTimport {8};
CToriginXYZ = CTimport {9};
```

Pixel size and slice spacing are required to resample the dose matrix to the same resolution as the reference image. For images exported from a TPS, the *SliceThickness* should be identical with the *SliceSpacing* since this is normally expected from a patient (CT) model for treatment planning purposes. *CToriginXYZ* is required to register the reference 3D image matrix with the dose 3D matrix.

Now we are ready to start importing and processing the dose distribution. Let's obtain the 3D dose matrix from a given DICOM file first, using the command sequence explained already (see the section 'Importing DICOM Data/Dose' earlier in this chapter):

```
RDdata = dicomread ('RDfileSample_1.dcm');
RDInfo = dicominfo ('RDfileSample_1.dcm');
dose3D = squeeze (double (RDdata) * RDInfo.DoseGridScaling);
```

As in the case of the reference CT image matrix, *RefImage*, the dose matrix resolution and location parameters are required:

```
RDPixelSpacing = RDInfo.PixelSpacing; RDoriginXYZ = RDInfo.
ImagePositionPatient.
```

The last parameter, *RDSliceSpacing*, is derived from the *RDInfo.GridFrameOffsetVector* containing a given slice offset from the first slice in millimeters. It is the same for a DICOM image export from a TPS, the dose slices (frames) should be exported with uniform spacing so the difference between any two neighbours should equal to the *RDSliceSpacing* value. Sometimes this is equal to the reference image slice spacing. This is when a TPS exports dose with the same native resolution as the reference (CT) image, so a given dose voxel size is, for example, [0.9766, 0.9766, 1.25]. For other TPSs, an isotropic voxel size is specified for dose calculation and is also used for the export, for example, [2.5, 2.5, 2.5]. So for a general situation, the imported dose matrix has to be resampled to match the reference image resolution as the first step of the registration process.

Returning to the problem, now we have the *dose3D* matrix with a voxel size of [*RDPixelSpacing(1)*, *RDPixelSpacing(2)*, *RDSliceSpacing*] which is required to resample to the reference (CT) image voxel size of [*CTPixelSpacing(1)*, *CTPixelSpacing(2)*, *CTSliceSpacing*].

Normally, the reference (CT) voxel size is equal or smaller than the *dose3D* matrix voxel size so the data resampling generally means interpolation in 3D. The *interp3* MATLAB in-built function is a natural option to do the job. Considering the variables introduced in this chapter so far, the appropriate command sequence is:

- Define the original dose matrix *meshgrid*

```
[X, Y, Z] = meshgrid (
RDPixelSpacing (2) * [0:1:size (dose3D, 2)—1],
RDPixelSpacing (1) * [0:1:size (dose3D, 1)—1],
RDSliceSpacing *[0:1:size (dose3D, 3)—1]);
```

- Define the new (interpolated) dose matrix *meshgrid*

```
[Xi, Yi, Zi] = meshgrid (
RDPixelSpacing(2)*[0: CTPixelSpacing(2)/RDPixelSpacing(2):
size(dose3D,2)-1],
RDPixelSpacing(1)*[0: CTPixelSpacing(1)/RDPixelSpacing(1):
size(dose3D,1)-1],
RDSliceSpacing*[0: CTSliceSpacing/RDSliceSpacing:
size(dose3D,3)-1]);
```

- Calculate 3D interpolation

```
doseInterp3 = interp3 (X, Y, Z, dose3D, Xi, Yi, Zi);
```

The respective *XYZ* dimension must match the corresponding matrix dimension (rows, columns, slices). Also note that the resulting matrix is also 3D with the voxel size equal to the voxel size of the reference (CT) image matrix.

Now, with the dose matrix interpolation ready, we can proceed to the final dose-reference (CT) image registration. This can be achieved by using the

```
>> RDoriginXYZ = RDInfo.ImagePositionPatient
```

parameter giving the *XYZ* coordinates of the upper left corner of the first slice (the first voxel transmitted). Knowing the reference (CT) image matrix dimensions and the first voxel's *XYZ*, *CToriginXYZ*, it is then easy to derive the reference (CT) image matrix indices *RCS* (row, columns, slice) corresponding to the location of the first dose voxel using the same equation as above (see the section 'Importing DICOM Data/Volumes of Interest' earlier in this chapter).

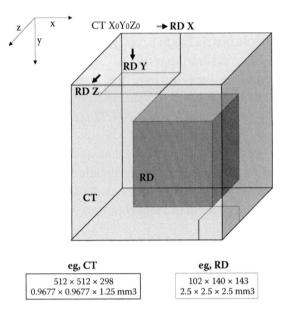

eg, CT	eg, RD
512 × 512 × 298	102 × 140 × 143
0.9677 × 0.9677 × 1.25 mm3	2.5 × 2.5 × 2.5 mm3

FIGURE 3.5   DICOM RT dose is generally a matrix of a different voxel size than the reference (CT) image matrix. The relative position of the dose matrix (RD) within the reference (CT) image matrix is determined by the difference of respective *ImagePositionPatient* attributes giving the *XYZ* coordinates of 'the first transmitted voxel'.

```
>> RDoriginRCS = round ((RDoriginXYZ - CToriginXYZ)./
[CTPixelSpacing(1), CTPixelSpacing(2), CTSliceSpacing]') + [1,1,1]'
```

To match the matrix dimensions with the IEC convention in use (IEC 61217) it might be necessary to flip the appropriate elements:

```
>> RDoriginRCS = [flipud (RDoriginRCS (1:2)); RDoriginRCS (3)].
```

The principle is demonstrated graphically in Figure 3.5.

By having the dose first voxel (reference image) matrix indices, it is then easy to create a dose matrix of the same dimensions as the reference image matrix with dose at appropriate locations. One can use the *match3Dto1st* script purposely written to perform this action and introduced in the section '3D Data/Matching 3D Data' in Chapter 2 (see the script available in Web Supplement 2.5).

```
>> dose3Dmatch = match3Dto1st (CT, RDoriginRCS, doseInterp3,
[1,1,1]);
>> dose3DregCT = dose3Dmatch {2};
```

Just as with RT structure masks, the dose match with the reference image can be verified visually by displaying isodose curves on the background reference (CT) image (*RefImage*), for example, for the slice 150:

```
>> figure; imagesc (RefImage (:,:, 150)); hold on
>> contour(squeeze(dose3DregCT(:,:, 150)), [.3,3] * max(dose3DregCT
(:))],'y','LineWidth',2)
```

displays the background CT slice in grayscale and the 30% (of the global dose maximum) isodose as a contour in yellow and specified line width.

Calculating and Comparing DVHs

In addition to verifying the reference image-RT structure-dose registration by means of visual comparison, there is another, possibly more efficient way: verifying dose-volume histograms. Assuming DVHs exported from a given TPS are correct, that is, that a given TPS was properly commissioned, a level of agreement between exported and calculated DVH is an efficient metric to assess the quality of a structure-dose registration.

Previously, we have demonstrated how to import a given structure DVH from the DICOM RTDOSE file by our own function script *RDdicom2dvh*. Let's proceed to calculating the DVH based on registered reference (CT) image-RT structure dose. Considering data available based on imports and processing demonstrated so far, the natural input for DVH calculation is a given structure mask in the form of the reference (CT) image matrix linear indices, and a 3D dose matrix registered with the reference (CT) image. Regarding the DVH output units, absolute values, both in dose and volume, are preferred for two reasons: i) having absolute dose matrix on the input and knowing voxel size is a natural and primary choice, ii) it is trivial to recalculate a given DVH to relative values whenever required (eg, for comparison with TPS-exported data). Regarding width and number of dose bins, there are many options. They can be determined such that, for example, there is always 100 bins covering the dose between the minimum and the maximum, or such that the dose bin width is fixed to, for example, 0.01 Gy, regardless of the dose range, and so on. Here is a possible algorithm for the MATLAB function.

Algorithm Scheme (Calculating DVH)

Function Syntax
    dose3D2dvh (dose3D, mask1D, voxelSizeMM)
Function Input/Output

    Input
        dose3D   Dose 3D matrix of the same size and resolution, registered with the reference (CT) image matrix. Dose units [Gy]
        mask1D   Sequence of linear indices of voxels forming a given structure within the reference (CT) image matrix
        voxelSizeMM   Voxel size of the registered dose3D matrix in millimeters [*rows, columns, slices*]
    Output
        dDVH   A 2D array of absolute dose [Gy] in the first column and corresponding structure-specific partial absolute volume [cm³] in the second

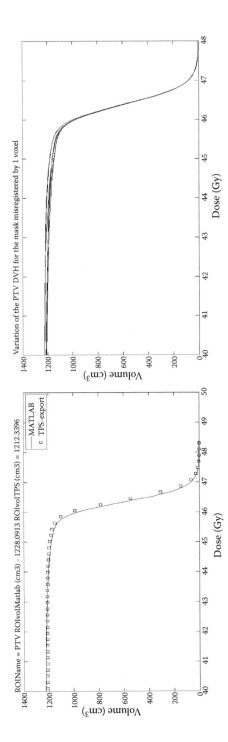

FIGURE 3.6 Comparing MATLAB-calculated DVH based on reconstructed DICOM CT, RTDOSE and RTSTRUCT data (solid line), and corresponding DVH directly exported from a commercial TPS (squares). The graph on the right shows the PTV DVH variation for the mask-dose misregistration by just 1 voxel for all 6 directions.

column. The number of rows equals the number of dose bins. Dose bin width is fixed and equals to 0.01Gy, so different structures have different DVH length.

Algorithm Steps

- Calculate the histogram for doses of all voxels addressed by the *mask1D* indices for the fixed vector of dose bin edges with the fixed 0.01Gy bin width

```
[voxelCount, doseBinEdges] = hist (dose3DregCT (mask1D),
[0:0.01:max (dose3D(:)])
```

- Organize the result in standard DVH format: column 1 = dose bins, column 2 = number of voxels

```
dDVH = [doseBinEdges', voxelCount'];
```

- Convert partial volumes from voxel count to cubic centimeters

```
dDVH (:, 2) = dDVH (:,2) * prod (voxelSizeMM)/1000;
```

By the intrinsic nature of the algorithm, a DVH produced by the *dose3D2dvh* function is differential. It is easily converted to the integral form using the own *diff2int* function (see the section '1D Data/DVH: Differential and Integral Format' in Chapter 2).

Having the calculated DVHs for selected structures, it is now possible to verify the reference (CT) image (RT structure) dose registration by comparing them with the corresponding exported DVHs. Comparing the MATLAB-calculated and the TPS-exported DVH pairs for one target volume is shown in Figure 3.6. Observed differences are due to how a given algorithm handles border voxels and due to residual misregistration, for example, because of rounding when converting from DICOM *XYZ* coordinates to matrix indices *RCS*. In the second part of the same figure, see the demonstration of DVH sensitivity to a structure mask vs. dose misregistration by just one voxel.

Using an absolute volume display reveals details about the difference in absolute volume and helps to understand whether the difference is due to either volume calculation or residual misregistration.

The absolute volume of RT structures reconstructed from a DICOM RTSTRUCT file may differ by a few percent even among commercial software, especially for small volumes. These differences indicate that the final outcome is not only about the shared data, that is, contour points, but also about how they are handled in particular software. Considering this experience, RT structures that are reconstructed by using simple algorithms presented in this chapter can be considered successful.

REFERENCES

ISO 12052:2017. (2017). Health informatics — Digital imaging and communication in medicine (DICOM) including workflow and data management. https://www.iso.org. Accessed May 11, 2017.

CyberKnife PEG. (2018). CyberKnife: Physics essentials guide. Accuray, Inc. P/N 029577A-ENG. CyberKnife. http://cyberknife.com

DicomIsEasy. (2016). http://dicomiseasy.blogspot.cz/. Accessed May 11, 2017.

DICOMLookup-A. (2017). DICOM Attribute requirement types. http://dicomlookup.com/type .asp. Accessed May 11, 2017.

DICOMLookup-B. (2017). DICOM SOP. http://dicomlookup.com/dicom-sop.asp. Accessed May 11, 2017.

DICOMLookup-C. (2017). http://dicomlookup.com/lookup.asp. Accessed May 11, 2017.

IEC 61217:2011 (2011). Radiotherapy equipment – Coordinates, movement and scales. International standard by International Electrotechnical Commission. https://webstore.iec.ch

MATLAB Help Offline R2013a. (2013). MATLAB, The MathWorks, Inc., Massachusetts, USA.

NEMA-A. (2017). About DICOM: Overview. http://dicom.nema.org/Dicom/about-DICOM.html. Accessed May 11, 2017.

NEMA-B. (2017). Radiotherapy modules. http://dicom.nema.org/dicom/2013/output/chtml/part 03/sect_C.8.html#sect_C.8.8.1.1. Accessed May 11, 2017.

NEMA-C. (2017). Treatment summary record module. http://dicom.nema.org/dicom/2013/output /chtml/part03/sect_C.8.html#sect_C.8.8.23. Accessed May 11, 2017.

NEMA-D. RT DVH Module. (2017). http://dicom.nema.org/dicom/2013/output/chtml/part03 /sect_C.8.html#sect_C.8.8.4. Accessed May 11, 2017.

CHAPTER 4

Modifying DICOM Data in Radiotherapy

LEARNING OUTCOMES

LO 4.1   Explain the rationale for DICOM data modification and the importance of doing this very carefully and responsibly

LO 4.2   Import and export DICOM CT, RTSTRUCT and RTPLAN files

LO 4.3   Compare differences in attributes and their values between the original and unmodified re-exported DICOM files. Recognize the importance of MATLAB DICOM dictionary

LO 4.4   Generate a modified CT image series and explain example applications from radiotherapy practice

LO 4.5   Modify DICOM RTPLAN file and explain example applications from radiotherapy practice

LO 4.6   Create own phantom (CT) image series

INTRODUCTION

This chapter extends the introduction to basic DICOM RT data by using a few examples. The task of data modification, together with the associated requirements necessary to do this safely and correctly, is the best approach to improve the understanding of the basic concepts of data structures and their application. Knowing the basic concepts of data organization, cross-referencing and the importance of key parameters can help significantly during troubleshooting. Sometimes it can actually solve a given problem itself. Of course, the key aspect is safety and nobody is encouraged to modify deliberately radiotherapy data intended to treat patients. However, data modification for purposes of investigation or troubleshooting is certainly justified, and it may actually contribute to safety through better understanding.

IMPORT AND RE-EXPORT WITHOUT MODIFICATION

In general, modifying a data set can be split into three steps: *original data import, data modification, modified data re-export*. Re-exported data is expected to be in the same format as the original. In Chapter 3 we presented some examples of importing and reconstructing basic RT data. In Chapter 1 we demonstrated MATLAB basic output and export options, but no example of exporting in DICOM format was presented. Therefore, let's have a look at the problem of import/re-export of four basic DICOM RT data into and from MATLAB, respectively.

DICOM Image Data

Images represent typical DICOM data. To import and re-export a given image series we can use the MATLAB inbuilt functions *dicominfo, dicomread* and *dicomwrite*. When you try the *dicomwrite* re-export on just one DICOM CT file (slice) using the following simple command sequence:

```
>> CTinfo = dicominfo ('CTsampleFile.dcm');
>> CTdata =  dicomread ('CTsampleFile.dcm');
>> dicomwrite (CTdata, 'testCTreExport.dcm', CTinfo);
>> CTinfoRE = dicominfo ('testCTreExport.dcm');
```

You can compare how the DICOM attributes and their values are converted during the process. You will find that while the DICOM study and series identifiers (*StudyInstanceUID, SeriesInstanceUID*) are preserved, the image/slice-specific identifier, *SOPInstanceUID*, changes as the *dicomwrite* generates a new identifier by default. It also generates MATLAB identification parameters regarding the creation software and other parameters related to the file (name, size) and data creation date and time. This means that the generated image series files are not identical and most likely will be recognized by a target software as a new image series, or new images of already existing image series, unless the series identifiers are modified manually. This is usually an advantage, as adding a new, potentially modified, image series is certainly a safer option than replacing an existing one, particularly when considering that there is usually a generous allowance regarding the number of series to import and process. An example of CT series data is available in Web Supplement 4.1.

When required, the basic parameters allowing easy and natural navigation among the existing and newly generated data can be modified manually before DICOM export. Table 4.1 shows a few examples.

Another important factor for the DICOM import-export is that some original parameters may import into MATLAB as structures named *Private_x_y*. These are DICOM attributes (DICOM tag *x, y* – see Chapter 3) not recognized by MATLAB at import, that is, they are not specified in the current MATLAB DICOM dictionary. The DICOM dictionary is the text file stored in the computer with the MATLAB installation under the path obtained by the following command: *DICTIONARY = dicomdict ('get')*. In principle, the DICOM dictionary may be expanded or modified to enable proper import and export of less common optional (type 3) attributes. This might be particularly important when these attributes are used by, for example, a commercial software for compatibility tests or to store other data important for

TABLE 4.1   Examples of basic series identification parameters making the new one, exported from MATLAB, easily distinguishable from the original and others

`SeriesInstanceUID = dicomuid;`	MATLAB generated new DICOM UID
`CTinfo.SeriesDescription = 'Modified';`	eg, specific series description
`CTinfo.SeriesNumber = 99;`	eg, specific series number
`CTinfo.StudyDate = '20170410';`	eg, specific study date
`CTinfo.SeriesDate = '20170410';`	eg, specific series date
`CTinfo.PatientName = 'xxxxxx';`	eg, specific patient name
`CTinfo.PatientID = '000000';`	... and ID

the given application. To handle the attributes properly, especially for the export, the DICOM dictionary might need expansion based on information available in the (eg, TPS) DICOM Conformance Statement of a given software.

An example of the image data import-modify-export algorithm is presented in the density override demonstration below. An example of the creation of an entirely new 3D image series is also demonstrated later in this chapter (see the section 'Creating Own Phantom Image Series').

RT Treatment Plan (RTPLAN)

Unlike dose or image data, the DICOM RT treatment plan standard does not include the 'image' data, just the 'metadata'. We can try a simple import and re-export using MATLAB inbuilt commands, for example, the command sequence:

```
>> origRPin = dicominfo ('RPsampleFile.dcm');
>> dicomwrite ([], 'testRPreExport.dcm', origRPin, 'CreateMode',
   'copy');
>> exportRPin = dicominfo ('testRPreExport.dcm');
```

to import sample treatment plan data (available in Web Supplement 4.1), re-export it *more or less* unmodified in another DICOM file and import it back in MATLAB again. Note the empty matrix variable (*[]*) to substitute general 'image' data that is expected for the primary intent of the MATLAB *dicomwrite* function. Also note the *'CreateMode'* parameter with its *'copy'* value. From the MATLAB's Help: unlike default value *'create'*, the *'copy'* simply copies all values from the input and does not generate missing values. Table 4.2 shows examples of differences in attributes present and their values for the sample DICOM RTPLAN file available in Web Supplement 4.1.

Some differences, such as those related to the file, for example, name, modification date and size, are expected. The others, such as *Width*, *Height*, *BitDepth* and *ColorType* and also attributes marked with (*) in Table 4.2 relate to (in this case imaginary) image data expected primarily for the *dicomwrite* function. But the difference in values of some attributes might be a problem when the exported data file is intended for replacing the original file, since in that case verification procedures applied in a given software to process the given data may disallow importing and processing of the modified data. For example, if *dicomwrite* generates a new *SOPInstanceUID*, there might be a problem when a new file (eg, RTPLAN) will be used together with the other original files (eg, RTDOSE, RTSTRUCT, CT) because of related data cross-referencing, the important safety feature. In such situations it depends on the particular

TABLE 4.2 Example comparing DICOM attributes and their value between the original DICOM RTPLAN file imported in MATLAB and the exported/re-imported again

Attributes with Different Values	Additional Attributes Present in the Exported/Re-Imported Data
FileName	*FileMetaInformationGroupLength*
FileModDate	*FileMetaInformationVersion*
FileSize	*MediaStorageSOPClassUID*
	MediaStorageSOPInstanceUID
Width ([] vs 0)	*TransferSyntaxUID*
Height ([] vs 0)	*ImplementationClassUID*
BitDepth ([] vs 0)	*ImplementationVersionName*
ColorType ('' vs 'grayscale')	(*) *SamplesPerPixel*
	(*) *PhotometricInterpretation*
SOPInstanceUID	(*) *Rows*
	(*) *Columns*
	(*) *BitsAllocated*
	(*) *BitsStored*
	(*) *HighBit*
	(*) *PixelRepresentation*
	(*) *SmallestImagePixelValue*
	(*) *LargestImagePixelValue*

Note: Some attributes have different values (left column) and some are present only in the exported/re-imported data set (right column). See the main text for the star-marked attributes.
(*) See the text.

DICOM implementation in terms of what cross-references are associated with, for example, *'warning'*, *'error'* or *'all clear'* status during the import and processing. Handling such situations is naturally complicated and very much case specific. A possible solution might be a manual post-export modification of problematic attributes using some DICOM editor or manual modification of the exporting process using *dicomwrite*. Neither solution is straightforward or elegant but an understanding of the basics and a few trial-error attempts may lead to a solution. See the section titled DICOM UIDs cross-referencing examples for additional information.

RT Structure Set (RTSTRUCT)

Similar to RT treatment plan data, the structure set too does not count with any 'image-like' data so the problem of importing and re-exporting unmodified data is analogous:

```
>> origRSin = dicominfo ('RSsampleFile.dcm');
>> dicomwrite ([], 'testRSreExport.dcm,' 'origRSin,' 'CreateMode,'
   'copy');
>> exportRSin = dicominfo ('testRSreExport.dcm');
```

Similar to the RT plan data, some observed differences can be expected when generating a new DICOM file and assumed image data as primary object for export using *dicomwrite* (see Table 4.3 for an example).

TABLE 4.3   Example comparing DICOM attributes and their value between the original
DICOM RTSTRUCT file imported in MATLAB and the same data exported/re-imported
again. (Some attributes have different value (left column) and some are present only in the
re-exported/re-imported data set (right column))

Attributes with Different Values	Additional Attributes Present in the Exported/ Re-Imported Data
FileName	*ImplementationVersionName*
FileModDate	(*)*SamplesPerPixel*
FileSize	(*)*PhotometricInterpretation*
FileMetaInformationGroupLength	(*)*Rows*
Width ([] vs 0)	(*)*Columns*
Height ([] vs 0)	(*)*BitsAllocated*
BitDepth ([] vs 0)	(*)*BitsStored*
ColorType ('' vs 'grayscale')	(*)*HighBit*
MediaStorageSOPInstanceUID	(*)*PixelRepresentation*
ImplementationClassUID	(*)*SmallestImagePixelValue*
SOPInstanceUID	(*)*LargestImagePixelValue*

Note: See the text for the star-marked attributes.
(*) See main text.

Although such a simple import-re-export procedure does not lead to a 100% identical DICOM file, *FileModDate* (file modification date) or DICOM UIDs (unique identifiers) naturally differ and some attributes may or may not be present in either version. Some may have zero value while empty in the other version, there is a good chance that the key data, that is, in case of the RTSTRUCT, the organization and record of coordinates of contour points forming a given structure VOI, are reproduced correctly. For overall complexity of background processes and possible dependence on the current version of MATLAB's DICOM handling functions, DICOM standard and DICOM conformance of a TPS or other software, some form of independent verification might be useful. For example, if only adding a structure or modifying one of few is required, parameters of all unmodified objects should remain unchanged, providing some form of confidence that the desired data modification is correct.

DICOM UIDs CROSS-REFERENCING EXAMPLES

As mentioned earlier, DICOM UID cross-referencing is one of the important safety features and ensures that only associated data is processed together. Natural, first-line attributes such as *PatientID*, are insufficient since each patient may have multiple courses, plans, structure sets, and so on, stored in the database. The RT structure set, RTSTRUCT, is firmly associated with the reference CT image so each contour set contains the exact UID reference to the associated reference CT slice as demonstrated earlier in Figure 3.4. One can also ask what potential the application has, for example, RTDOSE data, without a reference to the associated RTPLAN that actually produces the given dose distribution. Of course, there are applications when processing RTDOSE data regardless of the RT plan makes sense (see Chapter 5) but if, for example, there is an application requiring both data sets, it makes perfect sense to verify these two files for expected association. That is why, keeping to the example, the DICOM RTDOSE standard contains the attribute named *ReferencedRTPlanSequence* referring to the associated RT plan *SOPInstanceUID* so there is an option that any clinical software that checks this mutual association and generates a warning or error in the case of mismatch. What exactly is being verified and how is subject to a given software specific implementation and should be available to a user as DICOM Conformance Statement.

Let's demonstrate UID cross-referencing on a set of and RTPLAN specific DICOM data export: sample CT series, RTSTRUCT, RTPLAN and RTDOSE files are available in Web Supplement 4.2. Cross-referencing is presented by listing referencing and referenced UIDs in respective data sample type.

```
RTSTRUCT → CT

RTPLAN → RTSTRUCT

RTDOSE → RTPLAN
```

In addition to the cross-referencing listed in Tables 4.4–4.6, there might be other attributes linking associated data together. DICOM attributes may be used by processing

TABLE 4.4   DICOM RTSTRUCT attributes referencing to the associated CT image series within the sample data in Web Supplement 4.2

RTSTRUCT	CT
ReferencedFrameOfReferenceSequence.Item_1. FrameOfReferenceUID	FrameOfReferenceUID
RSinfo.StructureSetROISequence.Item_2. ReferencedFrameOfReferenceUID	FrameOfReferenceUID eg, for the structure item 2
RTReferencedStudySequence.Item_1.ReferencedSOPClassUID	SOPClassUID
ReferencedFrameOfReferenceSequence.Item_1. RTReferencedStudySequence.Item_1.ReferencedSOPInstanceUID	StudyInstanceUID
ReferencedFrameOfReferenceSequence.Item_1. RTReferencedStudySequence.Item_1. RTReferencedSeriesSequence.Item_1.SeriesInstanceUID	SeriesInstanceUID
ReferencedFrameOfReferenceSequence.Item_1. RTReferencedStudySequence.Item_1. RTReferencedSeriesSequence.Item_1.ContourImageSequence. Item_38.ReferencedSOPClassUID	SOPClassUID eg, for the CT image (slice) item 38
ReferencedFrameOfReferenceSequence.Item_1. RTReferencedStudySequence.Item_1. RTReferencedSeriesSequence.Item_1.ContourImageSequence. Item_38.ReferencedSOPInstanceUID	SOPInstanceUID eg, for the CT image (slice) item 38
ROIContourSequence.Item_9.ContourSequence.Item_6. ContourImageSequence.Item_1.ReferencedSOPClassUID	SOPClassUID eg, for the CT image (slice) associated with the structure volume item 9 and its contour sequence item 6
ROIContourSequence.Item_9.ContourSequence.Item_6. ContourImageSequence.Item_1.ReferencedSOPInstanceUID	SOPInstanceUID eg, for the CT image (slice) associated with the structure volume item 9 and its contour sequence item 6

TABLE 4.5   DICOM RTPLAN attributes referencing to the associated RTSTRUCT within the sample data in Web Supplement 4.2

RTPLAN	RTSTRUCT
ReferencedStructureSetSequence.Item_1. ReferencedSOPClassUID	SOPClassUID
ReferencedStructureSetSequence.Item_1. ReferencedSOPInstanceUID	SOPInstanceUID

TABLE 4.6   DICOM DOSE attributes referencing to associated RTPLAN within the sample data in Web Supplement 4.2

RTDOSE	RTPLAN
ReferencedRTPlanSequence.Item_1. ReferencedSOPClassUID	SOPClassUID
ReferencedRTPlanSequence.Item_1. ReferencedSOPInstanceUID	SOPInstanceUID

software for the data consistency checks. In case of the sample data in Web Supplement 4.2 these are *FrameOfReferenceUID* and *StudyUID* with identical values for all modalities exported. Identical *FrameOfReferenceUID* basically means using the same coordinate system throughout the subjected modalities (NEMA-A).

On understanding the basic concept of import/re-export DICOM RT data into MATLAB including cross-referencing aspects, we can try to proceed to practical examples of the data modification. One of the most common modifications everyone working in radiotherapy comes across is anonymization.

ANONYMIZING IMAGES AND OTHER DICOM DATA

There are numerous DICOM anonymizing software and even freeware, so this is not a particular application you might want to use MATLAB for, but data anonymization is an interesting field to demonstrate data modification.

There is the *dicomanon* inbuilt function in MATLAB for the purpose. It knows about all confidential medical information attributes and removes them. What attributes classify as 'confidential' is defined in (DICOM Supplement 55). The *dicomanon* function also specifies confidential attributes to keep. For more details see MATLAB Help.

EXAMPLES OF MODIFICATION OF DICOM RT DATA

Sometimes in clinical practice, there are non-standard situations. Some of them might be associated with errors interpreted by a given software such as data incompatibility. Some of these are 'real', that is, there is nothing else to do but obtain a new, compatible data set. However, sometimes a given problem is relatively small and fixing it manually can be both efficient and cost-effective. For example, sparing a patient from repeated CTs just because of insufficient nominal coverage or one missing slice outside a target area are probably examples of problems where an alternative solution is *worth considering*. Of course, safety first, and no data must ever be modified without being sure that a given modification does not change clinical data unintentionally, or brings any unexpected circumstances to a given clinical problem. This is a general problem of using bespoke software in radiotherapy (physics) with the key aspects being introduced in Chapter 1. In the next section, we will present a few practical examples where using the knowledge presented so far might be considered for fixing minor problems with radiotherapy data compactness, correctness, or compatibility.

CT Image Density Override

Ideally, a CT image represents a patient exactly as required for radiotherapy but for multiple reasons, this is not always the case. To help deal with some discrepancies, state of the art TPS support the *density override* function. In principle, this is a feature telling a given TPS that some voxels of a given CT model should have been assigned different physical or electron density (based on the relevant CT calibration curve) than would correspond to their true CT numbers from the original CT image. The most frequent situations requiring such correction include:

- Image artifacts caused by a high density object in the CT field of view causing very small or no X-ray attenuation resulting in incorrect image of areas where the affected X-rays were required for image reconstruction. These are typically shaded areas adjacent to metallic objects (tooth filling, prosthesis, pacemaker, fiducial markers, etc.) and streaking artifacts.

- Using a contrast agent or accessories required for desired image quality for planning and which will not be present during the actual treatment, eg, bladder catheter balloon and contrast agent alone. This includes the category of incompatible immobilization devices.

Correct, that is, representative, electron density is required for modern photon dose calculation algorithms to calculate dose distribution accurately. The standard approach is to contour a specific RT structure and assign a new CT number or physical or electron density. For example, objects that are present in the CT images but will not be present during treatment are assigned with density-specific parameters corresponding to air. In spite of the wide availability of this feature in modern TPS, there might be situations where having an option to produce an alternative CT series with modified CT numbers is useful. For example, when the CT density correction is forgotten or incorrect, a complex IMRT plan is prepared and a particular TPS does not allow changing density information without the necessity to reoptimize. In such an example situation, it might be useful to assess the potential impact of incorrect CT density on the given dose distribution by comparing the original plan dose with the same plan dose calculated on the corrected CT model. When the only chance to do this is to create a verification plan (normally calculated on a given phantom for IMRT plan verification measurements), then a corrected CT can be produced and uploaded as a new 'phantom' image set to obtain the dose distribution on the CT model which better represents a given patient during treatment. Based on the dose difference it can be decided whether the CT model correction and plan reoptimization is necessary.

To demonstrate how to produce an alternative image series with density override in MATLAB we use procedures presented in Chapter 3.

Algorithm Scheme

Function Syntax
  densityOverrideRS
Function Input/Output

  Prerequisite
    If possible or relevant, contour VOIs to override density in the TPS and export the CT and DICOM RTSTRUCT files. In MATLAB, navigate in the directory containing the DICOM CT series files and one DICOM RTSTRUCT file with a given structure set.
  Outcome
    Modified CT series files in a newly created folder.

Algorithm Steps

- Using the *importDCMseries*, *RSdicom2names* and *RSdicom2mask* functions import and reconstruct the CT model & binary masks for structures to density override as described previously (see the section 'Reconstructing 3D Dose Distribution' in Chapter 3)
  - Important from the export perspective is that as part of reconstructing the CT image series, the association of the CT 3D matrix slice rank with the DICOM file in the current folder is recorded in a specific variable (the *SliceLocationFile* variable within the *importDCMseries* function)
- For the given VOI(s) masks(s), change the CT values of the respective voxels in the CT 3D matrix as required
- Verify that the new CT 3D matrix is in the desired numeric format. When, eg, the data is type *double*, it might be required to convert it to appropriate type, eg, into 16-bit unsigned integer …

```
newCT = uint16 (newCT);
```

- Export the modified CT 3D matrix as a series of transversal CT images using the slice-specific metadata from the original CT series
  - Loop for the current folder with the original CT data files again
  - Record the slice-specific metadata using the *dicominfo* function
  - Consider the metadata modification as for examples given in Table 4.1
  - Using the recorded slice-location vs. DICOM file association stored in the *SliceLocationFiles* variable, find the corresponding slice in the modified CT 3D array to export, and …
  - … Export the given slice (*modifiedSlice*), with the associated *metadata* and under the new file name (*newFileName*), optionally in a new separate folder

```
dicomwrite (modifiedSlice, newFileName, metadata);
```

Example *densityOverrideRS* function is available in Web Supplement 4.3.

Modifying DICOM CT and RTSTRUCT

Modifying the DICOM CT series and associated RTSTRUCT data will be demonstrated by using a phantom CT series with three structures contoured (example data is available in Web Supplement 4.2). The example data come from a commercial TPS and contain the real CT image series. The *Volume1* structure corresponds to the object with density higher than the water equivalent phantom base, the *Volume2* structure is the air cavity. Let's generate a new data set with the density overridden in the *Volume2* region and with the *Volume1* shift 1cm towards patient right (HFS).

Algorithm Scheme

- Generate a new DICOM CT image series with the desired density override using the procedure *densityOverrideRS* described previously (see the section 'Examples of Modification of DICOM Data/CT Image Density Override' earlier in this chapter)
- Import the DICOM RTSTRUCT file in MATLAB using the *dicominfo* function
- Move the *Volume1* structure by 1cm patient right (HFS), ie, subtract 10 mm from all contour points defining the *Volume1* structure. See Table 4.7 for the key code details.
- Manage cross-referencing …
  - Although all seems ready to complete the job (with the new CT series exported already, and the modified *RSinfo* variable ready for the *dicomwrite*, at this point comes an important factor of cross-referencing.
  - From Table 4.4 we know the RTSTRUCT file contains cross-references to the associated CT slice *SOPInstanceUID* for each contour group stored. The new CT series generated using the *dicomwrite* function has the slices (files) associated *SOPInstanceUIDs* independent of the original values as the *dicomwrite* generates

TABLE 4.7   The structure array, *RSinfo*, with the *x*-coordinate of all contour points of all contour groups (*i*) for the structure *Item_2* (*Volume1*) adjusted by subtracting 10 mm (the *shiftContour* vector with every third element equal to –10)

```
RSinfo = dicominfo ('RSphantom.dcm');
for i = 1:length (fieldnames (RSinfo.ROIContourSequence.Item_2.
        ContourSequence))
    shiftContour = zeros (length (eval (['RSinfo.ROIContourSequence.
        Item_2. ContourSequence.Item_',num2str(i),'.
        ContourData'])),1);
    shiftContour (1:3:end,1) = -10;
    eval(['RSinfo.ROIContourSequence.Item_2.ContourSequence.
        Item_',num2str(i),'.ContourData=
        RSinfo.ROIContourSequence.Item_2.ContourSequence.
        Item_',num2str(i),'. ContourData + shiftContour']);
end
```

Note: Notice the application of *fieldnames* and *eval* MATLAB functions.

a new *SOPInstanceUID* by default. The current *RSinfo* contains the original *ReferencedSOPInstanceUID* for each given contour group, which would not be reset by the *dicomwrite* when creating a new RTSTRUCT file. So, as there is the new reference CT series available already, one might consider replacing the original *ReferencedSOPInstanceUIDs* with the new, current ones, as the easiest and natural solution. This is indeed possible, but requires knowing which pair of the original and new *SOPInstanceUIDs* correspond to the given CT slice to enable replacement. This can be done by listing the *SOPInstanceUIDs* of both CT series and sorting them based on some reference, eg, slice location coordinate, field rank in the folder (provided new file names are consistent with the old ones), etc.

- Next to the *ReferencedSOPInstanceUIDs*, the cross-linking job requires a few more DICOM attributes to link with the reference CT series (see Table 4.4) but the principle of updating is the same

• When the cross-referencing is done, the *RSinfo* is ready for the *dicomwrite* to generate the new DICOM RTSTRUCT file with the *Volume1* structure data modified and with the correct relevant referenced unique identifiers linking the file with the new, density overridden CT series files

```
dicomwrite ([], 'newRSwUIDs.dcm', RSinfo, 'CreateMode', 'copy');
```

The new, modified, CT and RTSTRUCT data can be tested in MATLAB simply by re-import and verification that the desired objectives have been met (see Figure 4.1), or, more interesting, they can be tested by importing into a dedicated commercial software.

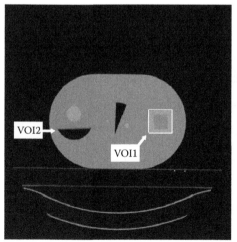 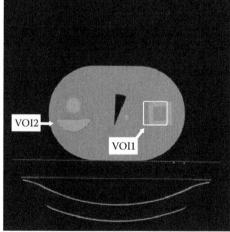

FIGURE 4.1 Example of DICOM CT and RTSTRUCT data modification: import in MATLAB (left)–modification-export–re-import (right) in MATLAB. Three VOIs contoured on the original phantom data. CT image files modified by assigning HU=300 in the volume of VOI2 (originally air). RTSTRUCT file modified by subtracting 10 mm from all contour points *x*-coordinates of the VOI1 (shift by 1 cm patient right).

Modifying DICOM RTPLAN

A useful application of modifying DICOM RTPLAN data is testing. For example, dose delivery using a dynamic multileaf collimator (MLC) for IMRT or VMAT is technically controlled by a sequence of control points spread across the delivery of a given treatment field (*ControlPointSequence* attribute). For IMRT, the basic parameters specified in a given control point are positions of all the leaves (*BeamLimitingDevicePositionSequence/ LeafJawPositions* attributes), and the corresponding fraction of total dose (MUs) to be delivered in the given radiation field (*CumulativeMetersetWeight* attribute). An example of the DICOM RTPLAN file related to a simple test 2-field IMRT treatment plan is available in Web Supplement 4.2.

Examples of tests efficiently addressed by DICOM RTPLAN modification include introducing a small error during the dynamic delivery by, for example, delaying a given leaf pair(s) at some control point(s). In principle, by doing this, one can test, for example, sensitivity of a detector array intended for routine pre-treatment plan verification. Letting the given TPS calculate the dose distribution associated with the original and modified treatment plan then allows possible clinical interpretation of the given error or its magnitude introduced by using clinical indicators such as DVHs. Based on such an assessment, it is then possible to think about establishing meaningful, or clinically relevant, tolerances for the technology. For example, what is the required accuracy of leaf positioning so that the difference between planned and delivered dose is clinically negligible? An example MATLAB code of the MLC leaf positioning modification by expanding a given moving gap between corresponding leaf pairs by 2 mm at each control point is shown in Table 4.8.

TABLE 4.8   Example of expanding a moving gap between paired leaves at each control point by accelerating leading leaves (No. 50-70) and delaying traveling leaves (No. 10-30) by 1 mm, respectively

```
for controlPoint = 2:138
    leafPos = eval (['RPorig.BeamSequence.Item_1.
        ControlPointSequence.Item_', num2str (controlPoint),'.
        BeamLimitingDevicePositionSequence.Item_1.
        LeafJawPositions']);
    leafPos (10:30) = leafPos (10:30) - 1;
    leafPos (50:70) = leafPos (50:70) + 1;
    eval (['RPorig.BeamSequence.Item_1.ControlPointSequence. Item_',
        num2str (controlPoint),'.BeamLimitingDevicePositionSequence.
        Item_1.LeafJawPositions = leafPos]);
end
```

CASE STUDIES: PROBLEMS SOLVED BY DATA MODIFICATION

Series 'Too Old'

Some commercial software might require only 'fresh' data to be used as the primary image set for dose calculation. This is entirely reasonable and it is the responsibility of the staff involved and, of course, a given radiation oncologist, to make sure the data for treatment planning is as representative as possible. However, sometimes the absolute condition that the data older than some limit is not importable at all might be too restrictive. Creating comparative plans or any other study are the examples. So, the DICOM CT image series (and study) date modification might be a quick solution to bypass such a restriction. Of course, there are many DICOM editors to do this job and one does not need MATLAB, but it is possible by using a simple image import-export procedure with appropriate attribute modification as illustrated in Table 4.1.

Missing Slice

The relatively simple problem of a missing single slice in the CT series may end up with the TPS rejecting its import for series inconsistency. Knowing how to import and re-export a DICOM CT series provides an option of fixing such problems by inserting a missing slice by interpolation. Considering what we know so far, the process is rather simple:

- Import and reconstruct a given CT series by using the *importDCMseries* function (see the section 'Importing DICOM Data' in Chapter 3)

- Identify a missing slice-location as a gap in the *SliceLocation* vector, the 1D array recording the file rank on import (in current folder) with the given slice location (sorted)

- Import DICOM CT images of two immediate neighbours of the missing location using the *dicomread* function

- Record full metadata on one of the neighbouring slices

  `metadata = dicominfo` (one of the neighbouring slice file name)

- Interpolate 2D between the neighbouring image arrays (see the section '2D Data' in Chapter 2)

- Copy the recorded metadata

  `modifiedMetadata = metadata;`

  and modify two parameters

- Generate new *SOPInstanceUID* using the MATLAB inbuilt function *dicomuid* (this will assure a unique identifier for the new slice within the series

```
modifiedMetadata.SOPInstanceUID = dicomuid;
```

- Modify the slice-location, ie, *ImagePositionPatient* vector so it corresponds to the slice-locations of the neighbours and slice-spacing parameters, eg, as follows

```
ImagePositionPatient_lower =
    metadata.ImagePositionPatient;
modifiedMetadata.ImagePositionPatient (3) =
    ImagePositionPatient_lower(3) + metadata.SliceThickness;
```

- Export the new CT image matrix including metadata to a new DICOM file

```
dicomwrite (newCTimgMatrix, 'newCTsliceFile.dcm',
modifiedMetadata)
```

Extending a CT Series

Insufficient CT coverage is rare but certainly a possible problem that may occur in any radiotherapy department. One of the basic rules of radiotherapy is to have the CT model long enough to provide realistic tissue data for all radiation beams. If coverage is short relative to field sizes and beam directions, especially non-coplanar, a problem might occur. As long as a missing CT coverage does not include the tumor to treat or a part of a critical organ where accurate DVH is essential, artificial extension of the CT series may be a reasonable alternative to requesting another planning CT. Of course, this is not an acceptable solution when the patient's body starts changing substantially just beyond the edge of the area originally covered. In that case, only a rescan would provide an adequate solution.

Another example from clinical practice relates to the minimum CT scan length required by a given TPS at the import even if this is only secondary or verification CT through the target area. When such a situation occurs, that is, the TPS requirement was not considered beforehand, extending the original CT series on both sides to meet the minimum scan length requirement might be an efficient solution without any compromises in the intended application.

In any of these practical examples, the solution using MATLAB and knowledge gained so far, is analogous to the previously presented missing slice problem. The only difference would be that instead of interpolating a new image within the original series, this time the extension images would be generated. Again, each newly generated image (file) must have appropriate DICOM attribute values:

- *SOPInstanceUID*

- *ImagePositionPatient*

- *SliceLocation*, and so on

Arithmetic Operations with Image Series

CTmip (*Maximum Intensity Projection*) and *CTave* (*Average*) are terms used in 4DCT to describe derivatives of the 4DCT acquisition. *CTmip* is typically used in lung cancer to guide the contouring of ITV (Internal Target Volume) and represents the whole range of expected tumor motion during the treatment. As suggested by its name, *CTmip* is calculated from a multiple image series as maximum voxel intensity over the series of otherwise equivalent 3DCTs acquired at a different respiratory phase. This approach ensures that any voxel associated with a certain position within the patient's body is visible, should the tumor have appeared in it at least once during the 4DCT acquisition period. Analogously, each voxel intensity of the *CTave* is calculated as the average voxel intensity over the series of 3D CTs acquired at various phases of the respiratory cycle. *CTave* is used instead of a single-phase or random-phase CT to better represent a free breathing patient at treatment. Although calculating 4DCT derivatives is generally included in commercial software constructing 4DCT, there may be situations in clinical practice when having an option, together with knowing how, can help to solve or investigate a particular problem. The following example illustrates this class of problems.

Sometimes, some respiratory motion management is expected to be beneficial and there is no 4DCT available. Depending on the 4DCT technology, it is also possible that a series of CTs at the opposite breath hold phase is preferred to 4DCT even when available. This can be due to residual motion artefacts, suboptimal breathing phase binning, and so on. If this happens, one can consider by using multiple CT series representing different phases of respiratory motion and calculating their own derivatives for purposes of treatment planning, for example, *CTmip* for ITV definition and/or *CTave* for more representative dose calculation. Of course, taking this approach enables additional options such as calculating *CTwAve*, that is, a weighted average, by giving higher importance to one of the input series, for example, 60% for expiration and 40% for inspiration, to reflect better reality expected on treatment. The algorithm is analogous to overriding density in a CT image presented earlier (please see 'CT Image Density Override'). The only difference is that the given numerical operation involves multiple CT data sets.

- Calculate the desired new CT 3D array using an appropriate numeric operation applied to the relevant 3D image arrays, eg, for 2 CTs available – one at expiration (*CTebh*), one at inspiration (*CTibh*) breath hold

```
CTmip = max (CTibh, CTebh);
CTave = (CTibh + CTebh)/2;
CTwAve = 0.4 * CTibh + 0.6 * CTebh;
```

Exporting the new CT images with the slice-specific metadata copied (and optionally modified) from one of the input CT image series, is equivalent to the procedures already described in the density overriding example.

Once again, it is important to be careful when using any modified data for actual clinical application. There are differences regarding safety even for applications with identical data. For example, producing an own *CTwAve* of the two breath hold CTs available, as presented in this chapter, for the purpose of dose calculation is relatively more questionable, compared with using the same product to 'just' verify that there is no significant difference in the dose distribution of the plan optimized on a single breath hold CT model. The first option means using the calculated image directly for the given clinical application, the second option is rather indirect since using the data provides 'only' *secondary information to consider*. However, in this particular case, it is relatively easy to verify the product independently just by comparing the source and the product image(s) in terms of Hounsfield units in subjected areas. So, in this particular case, the most challenging related aspect of using the *CTwAve* directly, ie, as the CT model for dose calculation, is probably the fact that the CT model is usually also used as the reference image for IGRT. And, quite obviously, the average image will be more or less blurred with direct impact on clinical matching at treatment. Considering that using yet another artificial image, or one of the originals, for IGRT only, shifts the whole thing from a practice orientation towards theory.

Dose as the Secondary Image

Another possible application of the same approach is the possibility of using the CT modifying feature to make some other, potentially important, information available. An example of such information is the area of previous treatment. Naturally, the easiest option to make this information available to a radiation oncologist deciding where to treat next, is to import the complete previous RT DICOM data (CT, RTPLAN, RTSTRUCT, RTDOSE) in a given TPS. However, importing DICOM RTPLAN and RTDOSE data from a treatment machine not commissioned in the local TPS is likely to be denied for safety reasons. In case the given department has a DICOM open software without such restriction, there is an option to let the doctor to use this but it may not be convenient. So, a possible option (using MATLAB) is to modify the previous RT CT data by 'burning' the isodoses from the dose distribution delivered directly into the image and export the product as an alternative secondary image series for the intended new treatment. Provided the secondary image (CT) is registered with the current CT model for treatment planning, the radiation oncologist can see the previous RT dose through standard viewing tools in the given TPS. Of course, doing so requires careful thinking before making the decision which isodose levels to 'burn' into the image and how to do it in a way that it does not corrupt the underlying CT image too much.

CREATING YOUR OWN PHANTOM IMAGE SERIES

Virtual or artificial phantoms are extremely useful in clinical radiotherapy physics, and also in imaging physics. There are many situations when manufacturing a real physical phantom is too demanding, too expensive, unnecessary, or simply imperfect to study a particular phenomenon. A common aspect of such situations is that it is a given phantom image series that is ultimately used for studying a given effect. So instead of expensively manufacturing a real physical phantom and scanning it to produce a desired image set, it is often, but not always, much more efficient to create a desired image series directly. The general advantage of virtual over physical phantoms is that they can be literary ideal with respect to the desired application. For example, they can have ideal dimensions and/or orientation according to voxel size or a task specific geometry, perfect CT numbers representing material composition, and so on.

The process of generating an own virtual phantom (CT) model, or image series, in DICOM format can be split in the following basic steps:

- Design a given voxel phantom as a 3D array using ROI specific numerical operations in MATLAB (see the section '2D Data' in Chapter 2 for examples)

- Create or copy associated metadata structure array

- Export both the image and metadata as a DICOM image series

Considering what we have learned so far, the middle step is probably the most complicated as creating an own metadata structure from nothing might not be trivial since it would be necessary to include all attributes required by the target software and to specify their values. The easiest approach to deal with this problem is to use a CT images series, preferably exported from a given target software, or a series which is accepted by a given target software with certainty, and apply it as a template to create an own phantom. After the CT series import and reconstruction in MATLAB, by using the *importDCMseries* function, it is trivial to zero the given CT matrix and create whatever phantom is preferred. Of course, the template series metadata key parameters must be verified when necessary. For example, attributes such as *PixelSpacing*, *SliceThickness*, *RescaleSlope*, *RescaleIntercept*, *Rows*, *Columns*, and so on, might be crucial for correct reconstruction of the phantom data in the given target software. An example of how to create an arbitrary 3D voxel phantom in MATLAB is presented in the next section.

Example CT Phantom: Preparing a 3D Voxel Phantom

As an example, let's create a virtual phantom CT image set importable into a TPS to study something. Let's set the phantom CT series objectives first:

- Series of standard 512×512 CT transversal slices with 1×1 mm^2 pixel size and 3 mm slice thickness, no gaps between slices, 101 slices in total corresponding to 303 mm longitudinal coverage

- Phantom roughly mimicking a human thorax with a body, lung and spinal bone with adequate CT numbers, eg, air and lung = −1000, body = 0, spinal bone = 1000

- Let's also add a patient couch as a rectangular shell structure: 3 mm body-equivalent density and air equivalent inside

A 3D array representing the above requirements can be created, for example, using the following sequence of MATLAB commands:

```
>> a2D = phantom ([1,0.75,0.5,0,0,0; 1,0.07,0.1,0,-0.3,0;
   -1,0.2,0.3,-0.3,0,0; -1,0.2,0.3,0.3,0,0], 512);
```

creating a composition of four ellipses differing in position, size and relative intensity value in order to match the phantom requirements for a transverse slice as a 2D base of the 3D matrix required. See MATLAB Help for explanations of parameters of the inbuilt *phantom* function.

Then 'draw' the patient couch shell as required:

```
>> a2D (385:387, 50:453) = 1; a2D (413:415, 50:453) = 1;

>> a2D (385:415, 50:53) = 1; a2D (385:415, 450:453) = 1;
```

Finally, copy the *a2D* base slice on all other 'thorax-mimicking' slices:

```
>> a3D = zeros (512,512,101); for i = 1:101 a3D (:,:, i) = a2D;
   end
```

At this point the matrix values vary among 0 (outside body, lung), 1 (body, couch shell), and 2 (spinal bone). To transfer the values into reasonable CT numbers in order to generate CT images, we can apply a simple linear conversion to match one of possible HU representation used in CT scanners and radiotherapy TPS.

```
>> a3Dhu = (a3D - 1) * 1000;
```

The *a3Dhu* array now represents the desired 3D voxel phantom of a very simplified human thorax. Selected orthogonal views are shown in Figure 4.2.

Example CT Phantom: Metadata Template and DICOM Export

Having a 3D voxel phantom ready as a MATLAB 3D array, we have to prepare a meaningful metadata structure array to accompany DICOM images for correct data reconstruction in a target software. Let us try to create a simple CT series DICOM export function that includes all major parameters for likely successful import into a target software.

Algorithm Scheme

Function Syntax
  exportCTseries (img3Dmatrix, dcmFilenameBase, voxelDim)
Function Input/Output

Input
  img3Dmatrix   3D matrix representing a desired 3D voxel phantom to export as DICOM CT image series
  dcmFilenameBase A base for names of generated files
  voxelDim   Desired voxel dimensions in millimeters *[PixelSpacing, Slice Thickness]*, eg, [1, 1, 3]

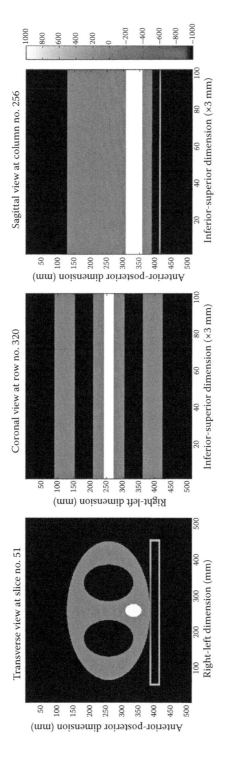

FIGURE 4.2 Major anatomical planes views of a simple human thorax 3D voxel phantom designed using sequences of simple commands in MATLAB including scale representing selected HU calibration for CT image series export.

Output

A new folder created with DICOM CT image series. Each file corresponding to one slice of the source *img3Dmatrix*.

Algorithm Steps

- Make sure the source *img3Dmatrix* is 16bit integer allowing negative values for the intended HU rescaling

```
img3Dmatrix = int16 (img3Dmatrix);
```

- Generate metadata general attributes that are common to all slices in the series

```
metadata.StudyInstanceUID = dicomuid;
metadata.SeriesInstanceUID = dicomuid;
metadata.FrameOfReferenceUID = dicomuid;
metadata.Modality = 'CT';
metadata.SeriesDescription = 'VirtualPhantom';
metadata.PixelSpacing = [voxelDim (1), voxelDim (2)];
metadata.SliceThickness = voxelDim (3);
metadata.ImageOrientationPatient = [1, 0, 0, 0, 1, 0]';
metadata.PatientPosition = 'HFS';
metadata.PatientName = 'xxxxxx';
metadata.PatientID = '000000';
metadata.RescaleSlope = 1;
metadata.RescaleIntercept = 0;
```

Note the arbitrarily decided fixed parameters including *ImageOrientationPatient* and *PatientOrientation* for this particular simple example DICOM CT series generating function.

- Generate slice and file-specific metadata and export
  - Create an export folder and navigate in it

```
mkdir ('CreatedPhantomCT'); cd CreatedPhantomCT
```

  - Loop for each slice of the input 3D matrix, generate slice-specific *ImagePositionPatient* and export

```
for i = 1:size(img3Dmatrix, 3)
    metadata.ImagePositionPatient = [-255, -255, (i -
    floor(size(img3Dmatrix, 3)/2)-1) * metadata.SliceThickness];
    dicomwrite (img3Dmatrix(:,:, i), [dcmFileNameBase,
    num2str(i),'.dcm'], metadata);
end
```

Note the *dicomwrite* function also generates default attributes unspecified by the *metadata* variable from the given example. For example, one of the most important is the *SOPInstanceUID* as a slice specific unique identifier. Although we did not specify this attribute in the algorithm/function above, the *dicomwrite* function automatically generates this mandatory parameter. Other attributes such as *Rows, Columns, Width, Height, BitDepth,* are all derived from the respective image data on the function input and also automatically generated without explicit specification in the *metadata* structure from the above example. On other hand, attributes such as voxel dimensions must be specified explicitly since there is no way the *dicomwrite* function derives the desired values. As in this example, the *dicomwrite* works slice-by-slice and we want all generated slices (files) to belong to one common study and series, these identifiers must be specified as attributes with common value in order to avoid the *dicomwrite* generating respective UID for each slice (file).

In the given example of the voxel phantom defined in the *a3Dhu* variable, the desired DICOM CT series is generated simply using the command:

```
>> exportCTseries (a3Dhu, 'phantomSlice_')
```

At first stage, correctness of the generated data can be verified by using the *importDCMseries* (see the section 'Importing DICOM Data' in Chapter 3). However, re-import in MATLAB is not, of course, the intended application. This is importing the series into a target software such as TPS as a new (phantom) CT series for testing.

REFERENCES

DICOM Supplement 55. (2002). Attribute level confidentiality (including de-identification). Accessed May 11, 2017.

MATLAB Help Offline. (2013). The MathWorks, Inc.

NEMA-A. (2017). Frame of reference UID. http://dicom.nema.org/medical/Dicom/2015a/output /chtml/part03/sect_C.7.4.html#sect_C.7.4.1.1.1. Accessed May 11, 2017.

CHAPTER 5

Simple 3D Plan Sum Using Rigid Registration

LEARNING OUTCOMES

LO 5.1   Explain the rationale for summing up treatment dose distributions in radiotherapy

LO 5.2   Explain the importance of image registration for summing up treatment plans

LO 5.3   Explain the limitations of rigid (6D) image registration for the purposes of plan sums

LO 5.4   Create an example platform in MATLAB in order to perform a simple rigid registration of two 3D images based on visual feedback

LO 5.5   Explain the transition from CT image registration to registration of doses and RT structure masks

LO 5.6   Explain the clinical objectives of the input RT plans relative weight factors before summing up the actual doses

LO 5.7   Explain the limitations of a simple dose-to-dose approach to summing RT doses

LO 5.8   Explain the importance of radiobiological modeling for summing up RT plans from different periods, with different modalities and different dose-fractionation schemes

LO 5.9   Explain the importance of plan sums reporting including the need for validation and approval of underlying image registrations and radiobiological modeling and choice of parameters

INTRODUCTION

This chapter presents a basic approach to summing two radiation treatment plans by using two different planning CT series. The operation of plan sum is often required in radiotherapy for situations where patients are undergoing multiple radiation treatments using various modalities (which are becoming more and more frequent). Plan sums on two different CT patient models (series) are not always supported by commercial treatment planning systems mainly because using a standard rigid registration of CT-dose structure depends on CT-to-CT image registration and its clinical relevance for a given problem. The plan sum concept is based on the assumption that rigid registration that is based on selected image objective (bony structures, fiducial markers or visible anatomy in subjected area) assures voxel-to-voxel matching making voxel dose sums clinically meaningful. Such an assumption is, of course, an ideal. For accurate and clinically correct dose sums it is required to map identical voxels on each CT series. This is a challenging task generally requiring non-rigid image registration as well as a reasonable anatomical model for the relevant organs. However, with a pair of contours for a given structure on each CT model, it is still useful to assess both DVHs corresponding to a pair of given structural volumes that represent 'two possible scenarios' of a given structure geometry relative to the sum of the delivered dose. In this chapter we will describe how to use basic CT-dose structure reconstruction that was introduced in Chapter 3 in order to perform a manual rigid registration based plan sum. Special attention will be given to discussion of alternative approaches to relevant data registration with regard to particular clinical objectives.

BASIC CONCEPT

By a 'plan sum' it is understood that this is the registering of two sets of CT-dose structures, each belonging to one treatment plan. Plan sums are often required to assess or estimate a given organ toxicity when new radiation treatment is indicated, especially in situations with particular or full dose overlap. In Chapter 3 it was demonstrated how to reconstruct a radiation treatment plan expressed as CT-dose structures registered data in MATLAB. Extending this knowledge to plan sums can be described by the following steps:

- Prerequisite: Let's have a *Plan1* and *Plan2* DICOM export of CT images, structures and dose (including DVHs when available) and place all files in a plan specific data folder. An example of the data is available in Web Supplement 5.1.

Then …

- Import and reconstruct CT-dose structures for *Plan1* using procedure described in Chapter 3

- Do the same for the *Plan2* data set

- Resample CT, dose and structure masks of *Plan2* to *Plan1*, or, resample both sets of data to arbitrary (isotropic) resolution, eg, $1 \times 1 \times 1$ mm^3

- Register CT matrix *Plan2* to match CT matrix *Plan1* using preferred objective and record the given rigid registration parameters

- Transform remaining *Plan2* data (dose and structure masks) using the same rigid registration parameters as saved in the previous step

- Verify *Plan2*, or both *Plan1* and *Plan2*, transforms by comparing MATLAB calculated and directly exported selected DVHs (when available)

- Now, with all *Plan1* and *Plan2* data registered, sum the dose with a given case specific plan weight factors. If relevant and if available, one can consider recalculating particular doses using radiobiological parameters as an alternative approach

RIGID REGISTRATION OF TWO CT SERIES

Although it sounds sufficient to describe the rigid registration, that is, 3D translation and 3D rotation, as a method for matching two plan data sets, it is actually useful to consider a few possible approaches specific to the problem.

Image to Image Registration

The first option is obvious and consists of matching the two underlying CT image sets represented by respective 3D matrices. Assuming identical matrix-to-patient orientation (ie, matrix dimensions and directions association with a given patient orientation) and spatial resolution, the problem is about finding a 6D transform matching the second matrix to the first one. This can be achieved either *automatically* by means of image pattern recognition (review, eg, AAPM TG132), or *manually*, using convenient visual feedback. In this section we will consider only manual match using the contour plot of the second image projected on the primary (CT) image displayed in grayscale. Alternatively, one can consider commonly used tools such as *split view*, *blend view*, and so on. The common reason for conducting manual matching using any visual feedback is to apply 3D translation and 3D rotation to the second matrix step by step to iteratively achieve the desired agreement. A simple approach based on only essential programming features is described by the following algorithm:

Algorithm Scheme (Manual Image-to-Image Registration Using the Contour Plot)

Function Syntax
  registerCT2toCT1 (CT1, CT2)
Function Input/Output

  Input
        CT1   Primary CT 3D matrix (Plan1)
        CT2   Secondary CT 3D matrix (Plan2) of the same voxel size as *CT1*
  Output
      CT2regCT1   *CT2* registered to *CT1* based on visual feedback and manual 6D rigid
                    registration

Algorithm Steps

- Define the central voxel as the initial reference voxel of the *CT1* and *CT2* matrices in order to select reference transverse, coronal and sagittal plane to guide the registration

```
refVoxelCT1 = floor (size (CT1)/2);
refVoxelCT2 = floor (size (CT2)/2)
```

- Define the initial CT window parameters [*CaxisLower CaxisUpper*] in order to display matching object (eg, bones) in the *CT1* images

Select the reference voxel CT1

- Loop for various *refVoxelCT1* until happy with the position of the reference orthogonal planes
  - Display reference transverse, coronal, and sagittal images for the current *refVoxelCT1* in MATLAB figure objects 1 to 3 and display current *refVoxelCT1* on-screen
  - Adjust the *refVoxelCT1* to select different reference planes in the next loop until satisfied

Select the optimal CT window

- Loop for various CT window levels until satisfied with the reference orthogonal images contrast
  - Display reference transverse, coronal, and sagittal images of the *CT1* in figures 1–3 and display the current window parameters [*CaxisLower CaxisUpper*] on-screen
  - Adjust the [*CaxisLower CaxisUpper*] to optimize the CT window until satisfied with contrast of the matching object (eg, bones)
- Keep figures 1–3 with the reference planes and optimal CT window displayed

Select the initial reference voxel CT2 and initial 3D rotation

- Loop for various *refVoxelCT2* until satisfied with the position of the reference orthogonal planes
  - Display reference transverse, coronal, and sagittal images for the current *refVoxelCT2* in MATLAB figure objects 4 to 6 and display current *refVoxelCT2* on-screen
  - Adjust the *refVoxelCT2* to select different reference planes in the next loop until satisfied
- Define an initial 3D rotation vector for *CT2* to match *CT1* as zero angles in degrees for transverse, coronal, and sagittal reference planes, respectively

```
rot3D = [0, 0, 0]
```

Select the contour plot values for the CT2

To visualize differences in the matching object in *CT2* on the *CT1* background we use the contour plot. The contours are specified by a user-defined *contourVector* of levels of the *CT2* image to display.

- Define the initial *contourVector* based on the matching object CT number read out from the *CT2* reference images in figures 4–6, eg,

```
contourVector = [0:200:400,800:300:1200];
```

- Hold figures 4–6 (now in optimized grayscale)
- Loop for various *contourVector* until satisfied with contour levels representing the matching object in *CT2*
  - Display reference transverse, coronal, and sagittal images for the current *refVoxelCT2* as green contour plots using current *contourVector* in figures 4–6 and display current *contourVector* values on-screen, for example, for the transverse plane

```
contour (CT2 (:,:, refVoxelCT2 (3)), contourVector, 'g')
```

  - Adjust the *contourVector* to select different levels to display in the next loop until satisfied

See Figure 2.7 for examples of contour plotting.

Fine match CT2 to CT1

At this point, figures 1–3 show the reference anatomical planes of *CT1* in optimized grayscale and figures 4–6 the reference anatomical planes of *CT2* in both the optimized grayscale and optimized green contours. We can close figures 4–6 now and proceed with matching *CT2* to *CT1* and displaying the *CT2* green contours on the *CT1* gray background for fine adjustment including 3D rotation:

- Close figures 4–6 and hold figures 1–3
- Loop for step-by-step iteration of the *refVoxelCT2* and *rot3D* in order to minimize visual differences in the matching object (eg, bones)
  - Match the *CT2* matrix to the *CT1* matrix based on current reference voxels (determining 3D translation) and current 3D rotation parameters using the 6D transformation function available in Web Supplement 5.2

```
match6Dto1st (refVoxelCT1, refVoxelCT2, rot3D)
```

  - Display reference transverse, coronal, and sagittal planes of the *CT1* defined by the *refVoxelCT1* in grayscale in figures 1–3
  - Hold figures 1–3
  - Display reference transverse, coronal, and sagittal plane for the matched *CT2* through the current *refVoxelCT2* as green contour plots using current *contourVector* in respective figures 1–3 and display the current values of the *refVoxelCT2* and *rot3D* on-screen
  - Adjust the *refVoxelCT2* and *rot3D* to match *CT2* to *CT1* in the next loop until satisfied
- Keep the final *CT2* matrix matched to *CT1* as the new *CT2regCT1* variable

Now, we have registered the *CT2* to the *CT1* matrix which also includes identical voxel size and matrix dimensions. The 6D transform matching the original *CT2* to *CT1* is determined by the *refVoxelCT1*, *refVoxelCT2* and *rot3D* vectors. The algorithm scheme is presented also graphically as a flowchart in Figure 5.1. Figure 5.2 then illustrates a practical

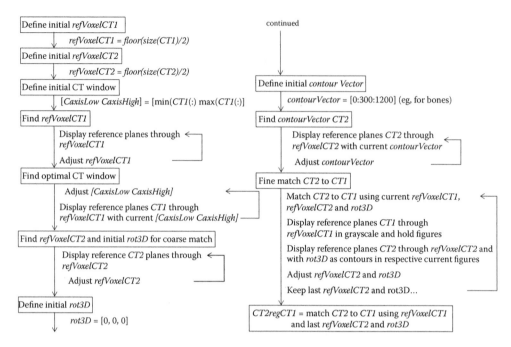

Define initial *refVoxelCT1*
  refVoxelCT1 = floor(size(CT1)/2)
Define initial *refVoxelCT2*
  refVoxelCT2 = floor(size(CT2)/2)
Define initial CT window
  [CaxisLow CaxisHigh] = [min(CT1(:) max(CT1(:)]
Find *refVoxelCT1*
  Display reference planes through ←
  refVoxelCT1
  Adjust *refVoxelCT1*
Find optimal CT window
  Adjust *[CaxisLow CaxisHigh]* ←
  Display reference planes *CT1* through
  refVoxelCT1 with current *[CaxisLow CaxisHigh]*
Find *refVoxelCT2* and initial *rot3D* for coarse match
  Display reference *CT2* planes through ←
  refVoxelCT2
  Adjust *refVoxelCT2*
Define initial *rot3D*
  rot3D = [0, 0, 0]

continued

Define initial *contour Vector*
  contourVector = [0:300:1200] (eg, for bones)
Find *contourVector CT2*
  Display reference planes *CT2* through ←
  refVoxelCT2 with current *contourVector*
  Adjust *contourVector*
Fine match *CT2* to *CT1*
  Match *CT2* to *CT1* using current *refVoxelCT1*, ←
  refVoxelCT2 and *rot3D*
  Display reference planes *CT1* through
  refVoxelCT1 in grayscale and hold figures
  Display reference planes *CT2* through *refVoxelCT2* and
  with *rot3D* as contours in respective current figures
  Adjust *refVoxelCT2* and *rot3D*
  Keep last *refVoxelCT2* and rot3D...
CT2regCT1 = match *CT2* to *CT1* using *refVoxelCT1*
and last *refVoxelCT2* and *rot3D*

FIGURE 5.1   Flowchart of the manual rigid match *CT2* to *CT1* using the *contour* plot feedback.

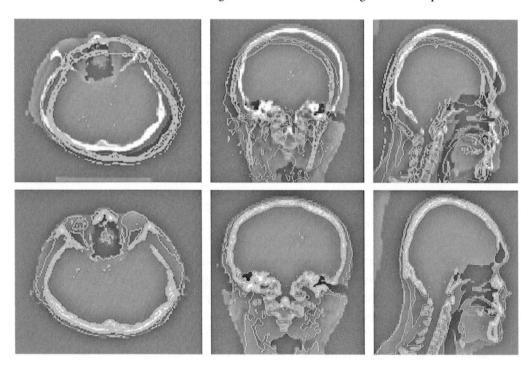

FIGURE 5.2   Transverse, coronal and sagittal views of the fixed *CT1* image through the selected reference voxel. The moving *CT2* image is represented by the green set of contours showing mostly bone interfaces. The first line of images shows the initial registration through central voxels of both CT series clearly indicating substantial mismatch in all directions. The second line of images demonstrates a reasonable 3D bone match.

application of the presented algorithm example. The example MATLAB script is available in Web Supplement 5.2.

Having structure masks and dose 3D matrices of *Plan2* registered to the *CT2* matrix requires only reproducing the same 6D transform to match also the structures and dose of *Plan2* with the *CT1* reference matrix. This will be shown later, but before then let's consider a few alternative options providing visual feedback to match two CT image matrices manually.

Registration Based on VOI Contours

One of many other options to provide visual feedback to drive manual rigid transform is displaying the contours of paired structures as contoured in both CT series to match. If, for example, a head-and-neck case has both left and right parotids contoured in both CT images, using the tools presented so far it is relatively simple to display the respective paired contours, for example, in reference orthogonal planes again, and guide the matching process using this particular paired structure or more paired structures when available.

From section on importing DICOM data in Chapter 3, we know how to reconstruct a given structure 3D mask in a matrix of the same size as the reference CT. Hence, the only technical difference to the algorithm described above is that a figure to assess the match between *CT2* and *CT1*, or better, *Plan2* and *Plan1*, is constructed in three instead of two steps:

- Display reference transverse, coronal, and sagittal planes of the *CT1* defined by the *refVoxelCT1* in grayscale in figures 1–3

- Hold figures 1–3

- Display reference transverse, coronal, and sagittal plane of a given structure mask 1 through the *refVoxelCT1* as a 1-contour plot (eg, white color)

- Display reference transverse, coronal, and sagittal planes of a given structure mask 2 through the *refVoxelCT2* as a 1-contour plot (eg, magenta color)

An example of a head-and-neck case is demonstrated in Figure 5.3. Note that sometimes the primary CT image (*CT1*) is referred as '*fixed*' while the secondary image to register with the primary is referred as '*moving*' image.

Although both versions of visual feedback presented so far are based on contour plotting, it is obvious that there is a substantial difference between them. In the first case, contours highlight contrast interfaces in a grayscale image which may or may not represent a clinically relevant matching object, whereas in the other case, a single contour represents the edge of a given structure mask, that is, an object having a direct link to DVH as a clinically relevant metric for the whole plan sum.

Registration Based on Pairs of Reference Points

An alternative or initial step to visual feedback guided manual registrations is using paired reference points (voxels) defined on each of the two image series to match. A minimum of

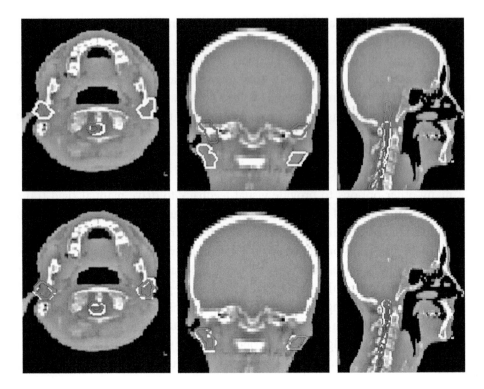

FIGURE 5.3   The transverse, coronal and sagittal views of the fixed *CT1* image through the selected reference voxel that shows both parotids in two planes and the spinal cord in two other planes as white contours. The moving *CT2* image is represented only by paired contours (both parotids and spinal cord) in magenta. The first line of figures shows initial registration through the central voxels of both CT models clearly indicating substantial mismatch mainly in the cranio-caudal direction. The second line of figures demonstrates a reasonable 3D match of both paired parotids following translation and –2° coronal rotation. Note the residual mismatch of the spinal cord contours originating from the non-reproducibility of contouring and possible deformation not covered by the manual rigid registration applied.

three non-collinear points define a plane with the origin representing the entire 3D array translation and rotation. If we find a minimum of three identical (anatomical) reference points in each 3D image, we can perform an automated point-based registration of the 3D arrays. Finding exact matching points may be a more difficult process than it sounds so some fine registration based on visual feedback might be required at the end. The underlying mathematics and appropriate MATLAB inbuilt functions are described in the following example.

Let's assume the following problem: There is a 3D array (*CT1*) representing a 3D image series, and there is another 3D array (*CT2*) representing another 3D image that we wish to register with the first array (*CT1*). A *3xN* array (*refP1*) contains *N* triplets of matrix indices of reference voxels of the array *CT1* and is similar for the same size *refP2* and *CT2*. Reference voxels rows, columns and slices form the first, second, and third line of the reference voxel matrix, respectively. The reference voxels should be as far apart as possible and in multiple layers of a given 3D array. The more reference voxels, the better the accuracy. Around 8 voxels is reasonable for the example demonstrated in this chapter. We are looking for a 6D transform matching the *CT2* 3D array with the *CT1* 3D array by finding the 6D transform matching the *refP2* voxels with the *refP1* voxels. Let us demonstrate a possible solution using the following function algorithm example.

Algorithm Scheme

Function Syntax
   match3Dpoints (CT2, refP2, refP1)
Prerequisites

   CT1 and *CT2* from the problem definition above have the same resolution. Both sets of reference voxels (*refP1*, *refP2*) are given in matrix voxel coordinates (rows, columns, slices).

Function Input/Output

   Input
         CT2   A 3D array to match with the *CT1* 3D array represented by the …
      refP1   Set of reference voxels (in the *CT1*)
      refP2   Set of reference voxels in the *CT2* array, representing identical (anatomical) positions as the *refP1* in the *CT1*.
   Output
      CT2toCT1   *CT2* 3D array registered with the target *CT1* array based on the 6D transform defined by two sets of reference voxels in respective 3D array

Algorithm Steps

- Expand both the *refP1* and *refP2* 2D arrays to prepare for matrix multiplication to perform special form of the affine transform. By transition from Cartesian coordinates to 'homogeneous' coordinates, any affine transform can be performed via matrix multiplication. The transition from Cartesian to 'homogeneous' is done by adding extra row of ones below the matrix of Cartesian coordinates (or by using the *cart2hom* inbuilt function in recent MATLAB releases), ie, for the *refP1*

$$refP1hom = \begin{pmatrix} refP1 \\ 1 \end{pmatrix} = \begin{pmatrix} r_1 & \cdots & r_N \\ c_1 & \cdots & c_N \\ s_1 & \cdots & s_N \\ 1 & \cdots & 1 \end{pmatrix}$$

where *r*, *c*, *s* stand for rows, columns and slices, and *N* is the number of reference voxels, and for the *refP2* analogically.
   Then, for the two sets of reference points, it stands

```
T * refP2hom = refP1hom
```

where T is the transform matrix 4×4 with the structure

$$T = \begin{pmatrix} A & d \\ 0 & 1 \end{pmatrix} = \begin{pmatrix} a_{11} & a_{12} & a_{13} & d_x \\ a_{21} & a_{22} & a_{23} & d_y \\ a_{31} & a_{32} & a_{33} & d_z \\ 0 & 0 & 0 & 1 \end{pmatrix}$$

Multiplying both sides of the previous equation by the transposed *refP2hom* matrix, that is,

```
T * (refP2hom * refP2hom') = refP1hom * refP2hom'
```

we obtain the square matrix *(refP2hom * refP2hom')* on the left equation side for which it is possible to find the inverse matrix *inv (refP2hom * refP2hom')* and multiply the equation with it again:

```
T * (refP2hom * refP2hom') * inv (refP2hom * refP2hom') =
(refP1hom * refP2hom') * inv (refP2hom * refP2hom')
```

The expression on the left side of the equation is the unity matrix so …
- … the required transform matrix T is finally obtained as

```
T = refP1hom * refP2hom' * (inv (refP2hom * refP2hom'));
```

- Due to rounding-off errors, the values in the last row of the transformation matrix T may not be 100% accurate so it might be better to replace them with accurate values *[0, 0, 0, 1]*. The MATLAB inbuilt function *tformarray* to be applied next is sensitive to this
- The final operation of the *CT2* matrix transposition using the transform matrix T is applied using the *tformarray* function:

```
R = makeresampler ('nearest', 'bound');
T1 = maketform ('affine', T');
CT2toCT1 = tformarray (CT2, T1, R, [1 2 3], [1 2 3], [a b c], [], []);
```

where R and *T1* are structures MATLAB creates when applying the *tformarray* function, *[a b c]* is size of the output 3D array (*CT2toCT1*) – usually the same as the input *CT2*.

Due to rounding-off errors, the images of the output array (*CT2toCT1*) might show some 'pepper and salt' type noise which is easily correctable using median filtering (see an example application of the *medfilt2* function in the section '2D Data/Filtering' in Chapter 2).

For details regarding functions applied see MATLAB Help. For more information about the mathematics see, for example, Foley et al. (1997).

Note the difference between the *match6Dto1st* (see the section 'Rigid Registration of Two CT Series/Image to Image Registration' earlier in this chapter) performing 3D translation and 3D rotation using manually entered parameters (reference voxels and 3 rotation angles), and the *match3Dpoints* function presented in this chapter, with the transform parameters derived from two sets of reference voxels on the input. Both function script examples are available in Web Supplement 5.2.

Testing the *match3Dpoints* function can be done on the sample phantom 3D CT series available in Web Supplement 5.3. Figure 5.4 shows the example of testing using eight reference voxels in total. The method is described in the figure caption.

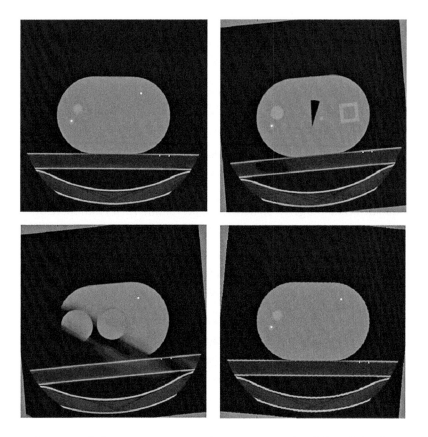

FIGURE 5.4 The original phantom *CT1* image/slice No. 120 with 2 (of total 8) reference 'markers' (area of $5 \times 5 \times 5$ voxels, HU = 1000) burned into the 'image' (top left). The phantom 3D image *CT2* after the 3D rotation (5°, 8° and 5° in transversal, coronal and sagittal plane, respectively)/slice No. 108 with one transformed reference marker visible (top right), and slice No. 133 with the second transformed marker visible (bottom left). Result of the *match3Dpoints (CT2, refP2, refP1)* operation: 3D array *CT2* (original *CT1* rotated 3D) transformed back to the original *CT1* based on the transform derived from two sets of the same reference points in the *CT1* and *CT2* 3D images – slice No. 120 (bottom right).

General Aspects

In general, rigid registration is insufficient for situations where there is deformation between the anatomy to match. In radiotherapy this happens often due to inter- and intra-fractional motion. Non-rigid or deformable image registration forms the class of approaches to address this problem. Notwithstanding the high level of sophistication, algorithms may have problems with the difference between the mathematical transform used and the physiological reality of the human body, including patient-to-patient individual variability. Assuming the DVH is the current standard describing the clinical effect of a given dose distribution on a given structure or a given target volume, the key assumption of clinically meaningful dose sums is *voxel-to-voxel mapping*. That is the ability to track each voxel of a given structure through the variations in position, shape, and volume it undergoes during the relevant stages of a radiotherapy regime(s). This is a challenging task from both a mathematical and clinical perspective. Using rigid registration only – as presented in this chapter – one cannot expect anything like this. However, considering the general deficit of accuracy in the whole concept, there are few aspects that can improve the overall robustness of even the simplest approach presented in this chapter:

- When there are more structures of interest and it is difficult to find a universal rigid transform to sufficiently match all of them, one can use multiple rigid transforms, each specific to a given structure.

- When there is a difference in structure contours for any reason (inter- and intra-observer variability, real changes in shape and volume, etc.), one can always use both structures to calculate a given plan sum DVH, and assess both together interpreting the result such that '*the given structure shape/volume varies and there are two samples in time to estimate potential clinical outcome*'.

- When a given structure shape is significantly isotropic and a structure-oriented plan sum is preferred, then one can consider finding the 3D rotation transform based on a solid anatomy such as spine and use a given structure just for translation.

Regardless of mathematical complexness and the level of sophistication of a given algorithm or approach, one should always be careful when interpreting results. One of the major objectives of this chapter is to let the reader realize that there are many assumptions behind every approach used in computer programs, some of which may compromise reality because of the general complexity of human anatomy and physiology.

Naturally, for visual feedback driven image registrations, the optimal *graphical user interface* (GUI) is a significant help. Using a GUI and interactive mouse control to go through manual iterations means a user-friendly form of the otherwise similar process as described in this chapter. However, incorporating a GUI not only significantly extends a given script length, but also it makes harder to follow the basic algorithm (see Figure 5.1). This is why a GUI has not been incorporated in given examples except of the simplest organization using 'docking' of multiple figure display as shown in Table 5.1. Using just a few commands allows figures to display in a more user friendly manner in order to facilitate guidance through the 3D registration process (see Figure 5.5).

TABLE 5.1   Comparing a command sequence for an example transverse view through the *ref Voxel* of the *CT1*: transverse (example), coronal and sagittal views in three separate windows, with axes displayed and figure number as a window identifier. The second option docks all three figures under three tabs in one figure window within the MATLAB default layout, with anatomy identifiers as tab names and no axis displayed

Basic Display Commands (eg)	More User-Friendly Display Commands (eg)
``` figure (1) imagesc (CT1(:, :, refVoxel (3))) caxis ([CaxisLower CaxisUpper]) colormap (gray) impixelinfo ```	``` figure (1) set (1, 'windowstyle', 'docked') set (1, 'name', 'transverse') set (1, 'NumberTitle', 'off') imagesc (CT1 (:, :, refVoxel (3))) caxis ([CaxisLower CaxisUpper]) colormap (gray) axis off impixelinfo ```

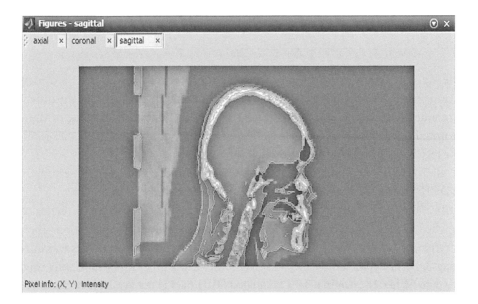

FIGURE 5.5   Example of docking figures to simplify iterative manual matching the 3D images.

In addition to the form of the visual feedback and details of the process design, user experience is a significant factor in the speed of obtaining the results and their accuracy. Although the process of image-to-image registration presented in this chapter is based on iterative visual assessment of a 6D transform controlled by two triplets of parameters inputted manually in each step may sound primitive, with some experience it enables results comparable with commercial solutions.

Although some TPSs by default export dose with the same spatial resolution as the primary CT, some allow optional choice of resolution, which may or may not reflect the underlying CT pixel and slice spacing. For two independent CT series, the pixel spacing is almost always different as it depends on the actual FOV which typically varies from exam to exam. Slice spacing depends on the imaging protocol applied so variation is less frequent but, of course, present, when multiple institutions, CT scanners, or protocols are involved.

These are factors influencing one important decision a user has to make: which RT plan/CT is to be the primary (*CT1*) and which the secondary (*CT2*) to be transformed to match the *CT1*? As image registration presented in this chapter is manual and driven by visual feedback, it is natural to prefer a finer resolution for the primary *CT1* to be used as the reference for all transforms of the *CT2*. Alternatively, one can consider a universal resampling to both subjected data sets to a uniform, for example, isotropic, fine $1 \times 1 \times 1$ mm³.

CO-REGISTRATION OF DOSE AND STRUCTURES – AND VERIFICATION

Assuming that the dose matrix and all relevant structure masks of *Plan2* have been co-registered with the patients's CT model (*CT2*), the next step in the plan sum process is to apply the 6D transform found in the *CT2* to *CT1* matching process to all related 3D matrices. This is to ensure that all objects of the problem (*dose1*, *structures1*, *CT2*, *dose2*, and *structures2*) are matched to the reference *CT1*. To apply the transform one can use exactly the same function as during the manual image-to-image matching process:

```
match6Dto1st (refVoxelCT1, refVoxelCT2, rot3D)
```

As the input matrix parameters are always identical (voxel size and matrix dimensions), the 6D transform can be applied straight away without having to consider the difference in pixel size and slice spacing and related angle nominal value when rotating in the sagittal or the coronal plane. Of course, this is irrelevant if all input 3D data has been resampled to isotropic (eg, $1 \times 1 \times 1$ mm³) resolution beforehand.

Once all 3D matrices are matched, it is then possible to proceed to 3D dose sums, isodose images, DVH plotting, and so on. However, before doing so there is one, potentially important step: verification of the *Plan2* data transform, that is, to what extent have the series of transforms applied to *CT2*, *dose2* and *structures2* impacted their mutual registration. The process of registration, or radiotherapy plan reconstruction, using comparison DVHs was introduced in Chapter 3. DVH data from the DICOM RD (dose) export file can be efficiently used as the reference data. When the original TPS calculated DVH data is not available or unsupported by a given TPS or not selected during export, one can consider calculating own reference DVH set after the initial plan reconstruction, that is, matching the dose and structures with the CT model. All registration transforms applied later can be validated at least against this reference.

DOSE SUMS AND REPORTING

Radiobiology Aspects

One of the most important parameters for plan sums in current commercial products with a similar objective (ie, plan sum with manual rigid registration), are relative weights of respective doses. These can be expressed in various ways:

- Relative weight factor with respect to actual exported (or saved) total dose data

- Number of fractions to actually apply from the exported/saved total dose data

- Total treatment dose to actually apply from the exported/saved total dose data

These are equivalent examples for simple situations such as summing up two or more sequential RT plans (eg, boost) with equivalent radiobiology.

Summing up RT doses from various courses with time interval of months or even years is another common practice to estimate the impact of the current plan on previously exposed healthy tissue. Compared with the previous simple category, for clinically meaningful interpretation this practice includes an additional complex factor – tissue repair. This means that it is not just what modality and what fractionation respective doses were used or are to be delivered that is important, but also to what extent the irradiated organs may be considered to have recovered from the previous radiation. A common clinical practice is to sum up doses 'as they are', that is, 'total Gy' and 'total Gy', look at the result and consider recovery only if the dose sum for a given organ exceeds the given tolerance criteria. When this happens, it is up to the radiation oncologist to consider the effects of tissue repair and adjust the respective plan related relative dose weight factor to better estimate the potential clinical impact.

More difficult situations occur when multiple modalities are in play, for example, brachytherapy with relatively high dose rates and generally completely different fractionation schemes than conventionally fractionated external beam photon RT, stereotactic hypofractionated RT, or hadron RT. In such situations the problem moves from exclusively physics to radiobiology and summing doses in terms of geometrical dose distribution becomes relatively less important compared with biologically effective dose concepts. Nevertheless, even for non-conventional fractionated external beam photon RT, the concept of DVHs is still a dominant quantitative parameter to describe normal tissue toxicity and tumor control. Therefore, plan sums in terms of geometrical dose distribution in total Gy delivered as presented in this chapter are, of course, relevant. However, one has to consider radiobiology as well. The dose (in Gy) as calculated by modern TPSs is still the major parameter determining the radiation effect on a tumor or an OAR. Following the principles described in this chapter, it is not difficult mathematically to consider a linear-quadratic (LQ) model and tumor and/or OAR specific radiobiology parameters to recalculate a given dose distribution to a distribution of derived quantities such as biologically effective dose (BED), and so on (eg, IAEA 2010). For each voxel we know the dose in Gy and its assignment to a given VOI with specific radiosensitivity and associated radiobiology parameters.

In modern RT it is not uncommon to deal with a problem of two or more RT courses of which one is stereotactic hypofractionated RT and another (two) conventional photon RT courses and vice versa. For further work on how to handle significantly different fractionation schemes, perhaps using the usual LQ model, the reader is recommended to consult the literature (Dale and Jones 2007).

A Few Thoughts on Reporting of Results

Let's assume now that after finishing the CT-dose-structure registrations for two or three RT plans, we have in general two or three doses plus one summation and usually two or three sets of structures. Some structures may be paired, that is, the same structures contoured on more than one CT model, some may be unique – plan specific. Considering only that rigid registration is applied, paired structures contoured on different CTs will differ even if VOI-based registration was used so it is reasonable to consider both (all) paired structures for clinical interpretation. Considering now that that this is a radiotherapy programming book mainly aimed at beginners, there are few simple and natural options on how to summarize and report the plan sum result in a brief, clear, and simple way without a complex GUI:

- Summed dose isodoses display

  - Including selected VOIs contours (?)

  - Only 3 main cuts (transverse, coronal, sagittal) through a reference voxel (isocenter (?), *Dmax* ?), or

  - Series of all transverse slices, possibly exported as, eg, TIF images so they don't lose too much quality as a result of the compression and can be displayed on any computer without special software (?)

- DVHs for each selected VOI (all available in case of paired structures)

- DVH stats summary report indicating structure specific V_{xGy} values, preferably selected by an operator creating the report based on clinical objectives (?)

An optimized form for presentation of results is an important quality aspect. It is important to consider the final user as well. If the latter is a radiation oncologist who needs to make a decision about the current fractionation and/or dose, simplicity and clarity of the presented data is just as important as the accuracy. A physician wants to make the correct decision based on correct data and he/she wants to do it *fast*. On the other hand, from a medical physicist perspective, it is equally important to make sure that a physician is fully aware of the assumptions, simplifications, and generalizations made during calculations. All major factors applied with an impact on the overall result must be considered when a decision is being made. With regard to the simple principles presented in this chapter, accuracy and preferences of the image-to-image registration are certainly key factors determining the final result presented in the form of plan sum DVHs. Therefore, if a sum DVH is presented to a physician, this should be done together with the review or even approval of the underlying image registration. The same applies to the choice of radiobiology parameters and/or model when this has to be included in summing the RT plans. For these reasons, in addition to the presentation of results described above, the example script available in Web Supplement 5.2 also includes:

- Image registration result in the form of screenshots of the last version of visual feedback applied

- Last verification DVHs, ie, comparison of the original TPS calculated DVHs vs. MATLAB calculated DVHs after all transforms have been applied

- Final rigid registration parameters to quickly reproduce the whole calculation at any time

TABLE 5.2 A table presentation of the example DVH stats for a given plan sum report summary. The main objective is to summarize the most relevant clinical parameters without having to read the data from DVH histograms

SpinalCord	V_{5Gy} [cm³]	V_{10Gy} [cm³]	V_{20Gy} [cm³]	D_{max} [Gy]	$D_{volMax2\%}$ [Gy]	D_{mean} [Gy]
DosePlan1+2	15.8	14.3	1.1	22.6	21.4	12.0
DosePlan3	5.0	0.0	0.0	8.0	6.9	3.2
DosePlanSum	16.2	14.6	9.8	30.2	27.6	15.1
	V_{5Gy} [%]	V_{10Gy} [%]	V_{20Gy} [%]	D_{max} [Gy]	$D_{volMax2\%}$ [Gy]	D_{mean} [Gy]
DosePlan1+2	72.5	65.3	5.2	22.6	21.4	12.0
DosePlan3	22.9	0.0	0.0	8.0	6.9	3.2
DosePlanSum	74.0	66.6	44.9	30.2	27.6	15.1

An example of the reporting of DVH stats for one selected structure is shown in Table 5.2. It demonstrates a few aspects of reporting quality:

- All stats presented for *Plan1*, *Plan2* and the plan sum

- Basic DVH stats relevant for every structure: (point) D_{max}, (robust dose maximum) $D_{volMax2\%}$ (see Table 2.5), D_{mean} in Gy

- Three structure specific V_{XGy} values selected either to demonstrate the difference or based on relevant toxicity criteria both in absolute [cm³] and relative [%] volume

REFERENCES

AAPM TG132. (2017). Use of image registration and fusion algorithms and techniques in radiotherapy: Report of the AAPM Radiation Therapy Committee Task Group No. 132. *Medical Physics* 44(7), e43–e76.

Dale R, Jones B. (2007). *Radiobiological modelling in radiation oncology*. United Kingdom: British Institute of Radiology.

Foley JD, van Dam A, Feiner SK, Hughes JF. (1997). *Computer graphics: Principles and practice in C*, 2nd edition. Boston, MA: Addison-Wesley.

IAEA. (2010). *Radiation biology: A handbook for teachers and students*. IAEA-TCS-42.

CHAPTER 6

Handling Regions and Volumes of Interest in Radiotherapy

LEARNING OUTCOMES

LO 6.1   Explain the main volume operations commonly used in contouring for radiotherapy planning including Boolean operations

LO 6.2   Expand and reduce volume shapes using the concept of anisotropic margin

LO 6.3   Explain the need for comparing volumes in radiotherapy

LO 6.4   Explain the concepts of (dose) conformity and (target) coverage

LO 6.5   Explain the basic concept of Internal Target Volume and the role of the merging volumes operation in both static and dynamic (tracking) coordinate systems

LO 6.6   Explain the assumptions for using surrogates of target volume position in radiotherapy and associated risk

---

INTRODUCTION

This chapter presents the rationale for volume handling in radiation treatment planning, review of volume definition, and basic operations including the application of Boolean operators. The major focus is an example algorithm for expanding and reducing any volume of any shape by a given anisotropic margin. An introduction to comparing volumes is driven by the presentation of the concepts of conformity and coverage applied mainly as quality indicators of dose distributions. The last sections of this chapter deal with the concept of internal target volume and the specifics for static and dynamic coordinate systems (tracking). An example algorithm assessing mutual bound of fiducial surrogates and target volume during breathing demonstrates a practical application of both the anisotropic margin and conformity-coverage function examples developed earlier.

DEFINING ROIs/VOIs – A REVIEW

The basic approaches to definition of ROIs (2D) and VOIs (3D) in MATLAB are introduced in the sections '2D Data/Regions of Interest' and '3D Data/Regions of Interest' in Chapter 2. How to reconstruct radiotherapy structures and VOIs, from DICOM CT and DICOM RTSTRUCT files created and exported from a specialist software such as a treatment planning system (TPS), is demonstrated in Chapter 3. The final outcome of any approach to the definition of ROIs can be a binary *mask* matrix of respective dimensionality (mostly 2D or 3D) with ones indicating pixels/voxels which otherwise belong to a given ROI and zeros. Having all ROIs in the form of binary masks enables essential operations by applying Boolean algebra.

BASIC ROI/VOI OPERATIONS

In this chapter we will introduce the basic VOI operations commonly applied when creating structures for radiation treatment planning. The operations can be categorized based on whether they apply to a single VOI only, or whether a second or higher number of VOIs are required to perform a given operation. Examples of operations commonly applied in radiotherapy contouring process are described.

Extending a VOI

Having a 3D binary mask of a given structure in the form of a 3D matrix with known rows, columns and slices associated with major anatomical planes and directions makes this operation trivial. Assuming a given VOI is contoured in the transversal planes corresponding to slices of a given 3D array, then extending the VOI contour, for example, by 5 slices in inferior direction (see Table 6.2 for matrix dimensions vs. anatomical directions) can be done simply by:

```
mask3D (:,:, infEnd + 1:infEnd + 5) = mask3D (:,:, infEnd * ones(1, 5))
```

where *infEnd* denotes the most inferior (non-zero) slice index of the given VOI mask.

Interpolation

Modern radiotherapy requires high spatial resolution in all dimensions including slice spacing. This is important, for the quality of IGRT reference images including Digitally Reconstructed Radiographs (DRRs). This represents relatively higher demands in the contouring process and having to contour a given structure on each CT slice is lengthy and inefficient. Hence, it is common to contour a selection of slices, preferably those where a given structure is well apparent and also in areas of significant change in shape from one slice to another. Contours on slices without direct contouring are then interpolated based on the contours defined directly. Interpolation has its pros and cons: a potential positive aspect is that a given VOI will be smooth in the interpolated area so that observer-related variations are suppressed; a potential negative aspect is the danger of insufficient sampling in areas of large changes in shape. So, a careful review of contours on all slices is necessary when a given VOI definition including interpolation is done. How to do interpolation, or better, resampling, in 3D is demonstrated in the section '3D Data/Interpolation and Resizing' in Chapter 2.

Margins and Wall Extraction

Sometimes it is useful to create a shell structure to represent a given structure wall. This is particularly useful for inverse planning, that is typically used in IMRT. Wall extraction from a given structure is a combination of applying symmetric margins (both to expand and reduce) and appropriate Boolean operations. Both categories are presented in separate sections below.

Boolean Operators

Together with adding margins, VOIs merging and subtracting represent essential tools for the radiotherapy contouring process. Basic graphical schemes, together with associated MATLAB logical operations applied to two given VOI masks, are demonstrated in Figure 6.1.

Clinical Examples

A few examples from routine clinical practice will help to associate the application of basic operations presented in this chapter with the real world.

Merging volumes is a typical operation for parallel organs such as kidneys and lung. Typically, both left and right organs are contoured separately and at the end a merged structure is created using the Boolean operation of *disjunction*, 'logical *OR*' (see Figure 6.1).

Another typical example of volume merging is, of course, a situation of multiple targets when it is usually convenient to define multiple, unconnected targets to be treated with the same dose as one joint structure (VOI).

In the frequent situation of overlapping target and critical (OAR) VOIs, VOI subtraction is used to create a complementary structure to ease the specification of optimization parameters in inverse planning. A typical example of such a situation is subtracting the rectum from the planning target volume (PTV) for a prostate case. Technically, this situation can be visualized as case 3 of Figure 6.1. Quite often an extra margin of, eg, 1 mm is added to the VOI being subtracted in order to create some separation between the original and resulting VOIs to account for expected dose gradient and to further ease the dose optimization process. Then, from a mathematical perspective, the operation extends to

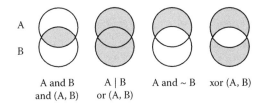

FIGURE 6.1 Venn diagrams of basic Boolean operations commonly applied in radiotherapy contouring, including related MATLAB logical operations applied to A and B VOI binary mask matrices.

adding an anisotropic margin to the VOI being subtracted before the subtraction itself. Another clinical situation when the operation of subtraction is useful is the Simultaneous Integrated Boost (SIB) with target sub-volumes to receive different doses.

Crop Structure

A VOI cropping operation technically equals the combination of basic operations. A typical example is when it is required to remove the extended part of a PTV extending outside the body, the adequate operation is the Boolean *conjunction* ('logical *AND*') between the PTV and the body structures. Again, it is common practice to create an extra gap between the adjusted PTV and the body which requires appropriate *reduction* of the body VOI before the conjunction operation.

Sample VOI masks of clinical structures to test Boolean operations are available in Web Supplement 6.1. As seen from the presented examples so far, the essential operation of volumes handling in radiotherapy is a given volume *anisotropic expansion* and/or *reduction*, so it deserves a whole section.

EXPANDING AND REDUCING ROIs/VOIs

Adding margins to VOIs is the essential component of radiotherapy planning. Typical examples include expanding target volumes GTV to CTV to account for 'subclinical involvement' of the tumor, and CTV to PTV to account for setup uncertainty (ICRU 83). A typical task sounds like this: *based on the given VOI construct a new VOI expanded by given amount in specific directions.* Although quite common, an isotropic margin cannot be considered sufficient to handle margins in general. For example, when the tumor is adjacent to a bone and there is clinical evidence that the tumor cannot expand into it, then it is clinically pointless to expand the given target isotropically to include the area. Also, margins to include, for example, uncertainty due to breathing motion might be asymmetric for the motion amplitude is usually larger in the superior-inferior direction than in the other two directions.

Unless an isotropic margin is required, treatment planning systems (TPS) typically offer the setting of the required margin parameters in millimeters in the major anatomical directions: anterior, posterior, left, right, superior and inferior. Constructing an isotropic margin is clearly a sub-problem of the general problem of anisotropic margin.

In general, margins do not necessarily mean volume expansion. It is well possible that a given VOI needs to be reduced by a certain amount in a given direction and this cannot be handled by a Boolean operation between two defined VOIs. In this chapter, a positive margin means volume *expansion* and a negative margin volume *reduction*.

Next to standard target volume expansions or reductions, other related applications include creating *shell* structures to help inverse planning. Computerized volume expansion and reduction is simply a very important feature that is applied in volume definition for treatment planning. It would be extremely laborious, and also inaccurate, to have to expand and reduce volumes by manual contouring. In this chapter, we demonstrate an example algorithm to create new VOIs by anisotropic expansion/reduction of the original volume.

The first question is to understand where in the process this particular task comes into play since the answer would impact the decisions that need to be taken regarding the input parameters of the example algorithm. As mentioned earlier, this is the phase of volume definition at the beginning of the treatment planning process. This means that a given patient CT model is already created, and some structures already contoured. Following the simple concept this book (particularly Chapter 3) uses, this means there is a 3D matrix representing the CT model and one or more 3D matrix masks of the same size representing the contoured volumes: voxels forming a given VOI have value 1, zeros otherwise. Hence, it is natural (at least for demonstration purposes in this chapter) to assume a given VOI 3D mask as the first input for a specific function created in MATLAB. In theory, one can consider also the DICOM RTSTRUCT form which is the list of contour points forming a given structure surface (see Figure 3.4), and achieving the given volume expansion/reduction by means of analytical geometry, but here we stick with the matrix form of representation. The next input parameter is, of course, a 6-element vector with instructions for how much it is required to expand or reduce the given volume. Provided this key input parameter is in millimeters, we need also voxel size information to relate matrix operations with real scale. The last input parameter required is patient orientation. Assuming a given VOI mask 3D matrix created in MATLAB, it is essential to relate matrix dimensions and directions to patient anatomy. In other words, we need to know what the matrix rows, columns and slices clinically represent. See Table 6.2 for an explanation of working frame applied in the presented example algorithm. Again, patient orientation is contained in the specific DICOM attribute *ImageOrientationPatient* available both in CT and RTSTRUCT files (see the section 'Importing DICOM Data/Volumes of Interest [RT Structures]' in Chapter 3), but for universality of the algorithm demonstration let's specify the patient orientation manually on the function input and also do the interpretation manually based on the information in Table 6.2. Considering all inputs, the function output form is clear: a new 3D mask matrix for expanded/reduced volume represented by an identical size 3D mask matrix as the function input.

Now for the most important thing: the algorithm itself. How will we do it? Considering basic RT data operations introduced in Chapters 1 and 2, and also the key aspect which is absolute independence of a VOI shape or dimensions (no square, rectangle, cuboid, sphere, ellipsoid, where dimensions can be easily altered analytically), let's use a very simple geometrical approach that consists in simulating a virtual motion. If we wish to expand a given VOI by a certain amount, let us 'move' it around its original position in a range of movement specified by the desired margin dimensions. Each 'visited' voxel is then added to a new volume. Assuming that in this way we produce a number of 3D masks with the original VOI at various positions, the desired outcome is obtained as 'logical *OR*', or *disjunction*, in Boolean algebra terminology. An example for an asymmetric horse-shoe shaped VOI anisotropic expansion specified by the vector [0,1,1,0,0,0], for desired expansion in major anatomical directions [A-nterior, P-osterior, L-eft, R-ight, S-uperior, I-nferior], that is, 'expand the given VOI 1 mm posteriorly and 1 mm (patient) left' for patient 'HFS' orientation, is demonstrated in Figure 6.2. For simplicity the voxel size is [1,1,1], that is, $1 \times 1 \times 1$ mm^3.

The opposite operation of a VOI reduction follows the same principle but applies a different logic at the end of procedure. A given VOI mask is moved around the original

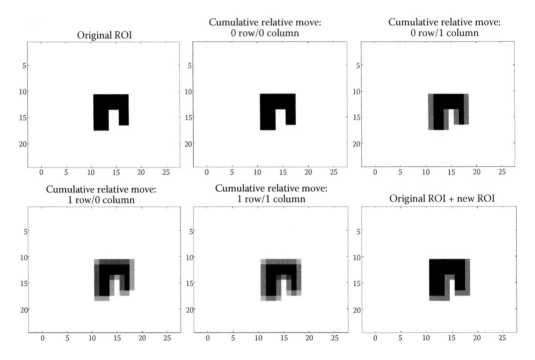

FIGURE 6.2 Explanation of the geometric concept of the algorithm using an example of asymmetric expansion: 1 voxel posterior, 1 voxel (patient) left, patient orientation 'HFS'. There are four possible combinations of the VOI position relative to the original. The respective cumulative relative motion figures show masks summed up consecutively with a given shift. All voxels with nonzero value form the output new VOI mask with the desired 1 voxel size expansion posteriorly and (patient) left.

position in the whole range of the desired reduction again but instead of 'logical *OR*' it is the 'logical *AND*' that applies. An example analogical to the expansion demonstrated in Figure 6.2 is presented in Figure 6.3.

None of the examples above would work correctly for a combination of expansion and reduction together. Due to the different logic that applies to cumulative VOI masks for the expansion and reduction case, using this example algorithm, it must split a combined case as an expansion followed by a reduction component. So, from the original input parameter [1,−1,−1,1,0,0], that is, 1 mm expansion anteriorly, 1 mm reduction posteriorly, 1 mm reduction (patient) left and 1 mm expansion (patient) right, we create the expansion-only component [1,0,0,1,0,0], perform the given expansion, and on the product we then perform the reduction-only component [0,−1,−1,0,0,0]. Zeros keep the irrelevant operation neutral with respect to a given component. In other words, we expand a given volume by what is required, and then we reduce the product by what is required. This example outcome is presented in Figure 6.4. Note that adding a margin in one direction, together with reduction by the same amount in the opposite direction, is equivalent to simple translation in a given direction by a given amount.

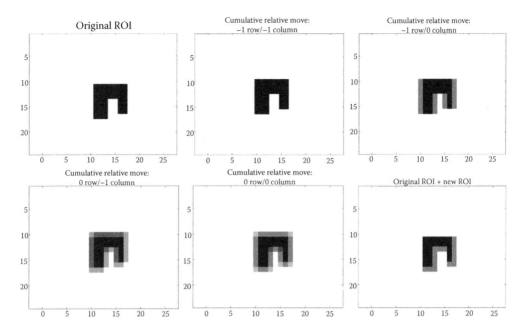

FIGURE 6.3 An example of a geometric concept applied to perform a VOI reduction by 1 voxel size in posterior and (patient) left direction for a patient at the 'HFS' orientation. The VOI is shifted in all four combinations (reduction specific) around the original position and the cumulative mask summed up consecutively. Only voxels with maximum value, that is, those valid for all combinations are selected for the outcome mask matrix. Note the relative moves are opposite to the analogical expansion example in Figure 6.2.

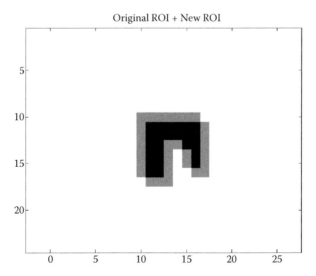

FIGURE 6.4 Example of combined anisotropic expansion and reduction. Combined anisotropic margin [1,–1,–1,1,0,0] = expansion-only [1,0,0,1,0,0] + reduction-only [0,–1,–1,0,0,0] of the product of expansion. Expansion anteriorly-right plus reduction by the same amount posteriorly-left equals to translation of the original VOI in the anterior-left direction.

Algorithm Scheme

Function Syntax

anisotropicMargin (mask3D, marginVector3Dmm, voxelSize, patientOrientation)

Function Input/Output

Input

mask3D	A 3D array (matrix) mask for a given VOI. Ones at voxels within a given VOI, zeros otherwise
marginVector3Dmm	Instructions for anisotropic margin calculation. In millimeters. For six major anatomical directions: anterior (A), posterior (P), left (L), right (R), superior (S), inferior (I): *[A, P, L, R, S, I]*. Positive value for VOI expansion, negative for VOI reduction, in a given direction.
voxelSize	Voxel dimensions in [mm] for converting the desired margin from millimeters to voxels
patientOrientation	A string parameter giving patient (*mask3D*) orientation, that is, linking matrix directions (rows, columns, slices) with anatomical directions and respective margin parameter: *'HFS'*, *'HFP'*, *'FFS'*, *'FFP'* (see the section 'Importing DICOM Data/ Image Series' in Chapter 3)

Output

mask3Dnew	New 3D mask array for a given VOI with margins added/ subtracted

Algorithm Steps

- Determine the *mask3D* reference point for relative 'motion' around it

```
refVoxel = floor (size (mask3D)/2);
```

- Translate the *patientOrientation, voxelSize* and *marginVector3Dmm* parameter in the 6-element vector (*marginVector3Dvoxels)* of the VOI shift in voxels in the respective dimension. To distinguish different anatomy-matrix direction assignments for four possible patient orientations, simply use the *switch* command. See Table 6.1 for a demonstration of example code and Table 6.2 for a summary of 4 major patient orientations and anatomy vs. 3D matrix coordinates and directions bounds to explain the signs vectors applied in this part of the script
- Split the *marginVector3Dvoxels* in two based on the original *marginVector3Dmm* using the following criteria:
  - All positive elements of the *marginVector3Dmm* mean the VOI expansion in respective direction so to separate the expansion from the reduction to apply different logics to different vectors can replace the original one

```
marginVector3Dvoxels_EXPAND = marginVector3Dvoxels;
marginVector3Dvoxels_EXPAND (marginVector3Dmm < 0) = 0;
```

Note: Zeros for the elements indicate the VOI reduction.

TABLE 6.1   MATLAB script to convert anisotropic margin parameter from millimeters *marginVector3Dmm* = [A, P, L, R, S, I] to voxels *marginVector3Dmm* for four major patient orientations

```
switch patientOrientation
    case 'HFS'
        marginVector3Dvoxels =[[A,P]/voxelSize(1),
        [L,R]/voxelSize(2), [S,I]/voxelSize(3)].*[-1, 1, 1, -1, -1,
        1];
    case 'FFS'
        marginVector3Dvoxels = [[A,P]/voxelSize(1),
        [L,R]/voxelSize(2), [S,I]/voxelSize(3)].*[-1, 1, -1, 1, 1,
        -1];
    case 'HFP'
        marginVector3Dvoxels = [[A,P]/voxelSize(1),
        [L,R]/voxelSize(2), [S,I]/voxelSize(3)].*[1, -1, -1, 1, -1,
        1];
    case 'FFP'
        marginVector3Dvoxels = [[A,P]/voxelSize(1),
        [L,R]/voxelSize(2), [S,I]/voxelSize(3)].*[1, -1, 1, -1, -1,
        1];
end
```

TABLE 6.2   Patient orientation vs. matrix coordinates and directions for (CT) models and VOI masks reconstructed using the *importDCMseries* and *RSdicom2mask* function examples that were presented in Chapter 3: Anterior (A), posterior (P), left (L), right (R), superior (S), inferior (I)

CT Matrix Coordinate	CT Matrix Direction	Main Patient Orientations and Respective Anatomy Direction			
		HFS	FFS	HFP	FFP
$CT(i, :, :)$	From row 1 to last	P	P	A	A
$CT(:, j, :)$	From column 1 to last	L	P	P	L
$CT(:, :, k)$	From slice 1 to last	I	S	I	S

```
marginVector3Dvoxels_REDUCE = marginVector3Dmm;
marginVector3Dvoxels_REDUCE (marginVector3Dmm > 0) = 0;
```

Note: Zeros for the elements indicate the VOI expansion.

(Now calculating the VOI expansion component …)

- Loop for all combinations of the VOI relative shift given by the *marginVector3Dvoxels_EXPAND* to 'simulate all possible VOI locations to fill desired margin
  - For each combination *[i, j, k]* of relative shift in row, column and slice determine the VOI mask for the given 3D shift using the *match3Dto1st* function (see the section '3D Data/Matching 3D Data' in Chapter 2) and add up this mask consecutively in the *mask3DconsSum* variable

```
movedVOImask = match3Dto1st (mask3D, refVoxel, mask3D, refVoxel -
                [i, j, k]);
mask3DconsSum = mask3DconsSum + movedVOImask {2};
```

Note: The product of the *match3Dto1st* function keeps the new second matrix in the second output element (cell array)

- In the sum of VOI masks (*mask3DconsSum*) with all relevant shifts included, find all non-zero voxels. These represent positions where the VOI appeared at least once during the simulated motion chosen as a simple algorithm for adding margin in this example:

```
mask3Dexpand = zeros (size (mask3D));
mask3Dexpand (mask3DconsSum > 0) = 1;
```

(Now calculating the VOI reduction component on the expanded component product …)

- Loop for all combinations of the VOI relative shift given by the *marginVector3Dvoxels_REDUCE* to 'simulate all possible VOI locations to fill the desired margin
    - For each combination *[i, j, k]* of relative shift in row, column and slice determine the VOI mask for the given 3D shift using the *match3Dto1st* function (see the section '3D Data/Matching 3D Data' in Chapter 2) and add up this mask consecutively in the *mask3DconsSum* variable:

```
movedVOImask = match3Dto1st (mask3D, refVoxel, mask3D, refVoxel -
[i, j, k]);
mask3DconsSum = mask3DconsSum + movedVOImask {2};
```

- In the sum of VOI masks (*mask3DconsSum*) with all relevant shifts included, find all voxels with maximum intensity, ie, those voxels which are present in all masks representing each relative shift

```
mask3Dfinal = zeros (size (mask3D));
mask3Dfinal (mask3DconsSum == max (mask3DconsSum (:)) = 1;
```

An example MATLAB script of the *anisotropicMargin* function following the above algorithm is available in Web Supplement 6.2.

COMPARING ROIs/VOIs

In radiotherapy, measurement of volumes plays a significant role not only to assess temporal changes such as tumor progression or reduction, tumor volumes are also used as one of the decision parameters about type of (radio)therapy, prescribed dose, and so on. Working with a CT model 3D array and radiotherapy structure 3D masks as presented in this book, make determining volume of a given structure (*VOIvol*) trivial. Assuming the voxel size given as a 3-element vector for CT pixel size and slice spacing then:

```
VOIvol = length (mask3D == 1) * prod (voxelSize)
```

in the specified units of the voxel dimensions.

Conformity Index and Coverage

However, rather than measuring structure volumes, from radiotherapy physics and particularly treatment planning perspective, a very interesting problem is assessing the conformity of dose distribution. Provided a given target volume receives the desired treatment dose, a given treatment plan conformity is quantified by the volume of irradiated normal tissue. The lower the volume of normal tissues irradiated, the better the treatment plan quality. It is quite natural that the dose optimization process, regardless of whether computerized (inverse planning) or manual (forward planning), focuses mostly on high (treatment) isodose in order to minimize the exposure of normal tissues to high doses. Volumes with lower isodoses are often controlled with relatively less success since they are dominated by factors either excluded or only partially included in the dose optimization process (eg, type of radiation, energy, beam geometry). So, it is not a big surprise that dose conformity traditionally has been associated with the target volume vs. treatment isodose volume relationship. Having said this, the simplest definition of a quantitative parameter reflecting conformity of a given dose distribution to a given target, the *conformity index*, is a simple ratio of the two volumes:

$$CI = \frac{V_{XGy}}{V^{PTV}}$$

where V_{XGy} represents the volume of a given treatment isodose and V^{PTV} the volume of the PTV. The major disadvantage of such a definition is obvious; it does not say anything about the mutual relationship in terms of the relative position between the two volumes, that is, it does not reflect target *coverage* by treatment isodose at all.

Probably the most used formula for the conformity index today differs by replacing the treatment isodose volume in the denominator by the volume of the PTV covered by the treatment isodose, that is,

$$CI = \frac{V_{XGy}}{V_{XGy}^{PTV}}$$

Such a definition of the *CI* may give an infinite value when the two volumes do not overlap. Figure 6.5 demonstrates three possible relationships between the two volumes and the respective *CI* values. From the figure and the formula one can see that even the improved definition of *CI* does not cover all clinically relevant aspects. In this case, it does not sufficiently handle the coverage of a given PTV itself. Therefore, to quantify treatment isodose conformity and to avoid neglecting an important clinical parameter, one should use both the *CI* and the *COV* (coverage) parameters together. A combination of product of both can also be considered as a reasonable alternative approach in obtaining a single value instead of two, but probably without a significant gain in terms of the sensitivity to both variables involved. More comprehensive reviews of other definitions of conformity parameters can be found in the literature (eg, Ohtakara et al. 2012).

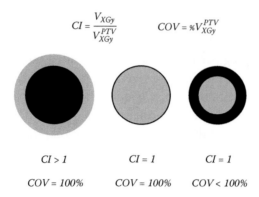

FIGURE 6.5 The conformity index (*CI*) including the PTV coverage and the coverage (*COV*) parameter demonstration for three scenarios: treatment isodose (gray), fully encompassing a given PTV (black), treatment isodose exactly matching the PTV, and treatment isodose volume partially covering the PTV.

From the *CI* and *COV* definitions it is clear that one can obtain the necessary input data from the DVHs. The *COV* can be read directly from a given PTV DVH in an integral form with a relative volume scale as the % V_{XGy}^{PTV}. The V_{XGy}^{PTV} parameter is in absolute volume units in order to match the units of the given isodose volume, V_{XGy}, which can be read from the DVH of any structure encompassing fully the given *XGy* isodose (eg, the body or skin). Again, to ensure the same units in both elements of the fraction, the volume must be made absolute in cm³.

Following the concept of this book, it is convenient to consider determining conformity parameters in a general sense, based on 3D mask matrices of two general volumes. This can also be useful for applications other than dose conformity as will be demonstrated later. An example of a simple MATLAB script to calculate both the *CI* and *COV* for 3D masks of two general volumes is demonstrated in Table 6.3. The script is also available in Web Supplement 6.3. Modifying the MATLAB script to alternatively calculate defined *CI* from the literature (eg, Ohtakara et al. 2012) is a trivial combination of the two parameters determined in the demonstrated example.

TABLE 6.3   Example of a MATLAB script used to calculate both the *CI* and the *COV* parameters for 3D mask matrices of two general volumes

```
function [CI, COV] = confParams (mask3D_1, mask3D_2);
% two masks in the same size 3D matrices with the same
% pixel size and slice spacing
% mask3D_1 = reference
% mask3D_2 supposed to cover mask3D_1

maskSum = mask3D_1 + mask3D_2;

CI = 1 + length (find (maskSum == 1))/length (find
    (mask3D_1 == 1));

COV = length (find (maskSum == 2))/length (find
    (mask3D_1 == 1)) * 100;
```

INTERNAL TARGET VOLUME

As per the ICRU 62 definition, the internal target volume (ITV), or associated internal margin, represents the known or estimated internal motion of the clinical target volume due to the physiological processes in patient's body such as breathing. Having a CTV contoured on CT images representing differentiated breathing phases (inspiration and expiration) allows for both CTVs to merge to create a new VOI:

$$ITV = CTV_{ins} + CTV_{exp}$$

Of course, sampling of two 3D images only can be considered insufficient for complex internal motion due to breathing. Additionally, representativeness of a recorded image, that is, to what extent a given image represents well the expected patient performance during dose delivery, can also be an issue when such an approach is used clinically. With 4DCT one can acquire more data and more samples during the breathing cycle; for example, ten 3DCTs can be acquired, each representing a different phase of the range of breathing. One might expect having a full breathing cycle covered by multiple 3D CT image series (ie, 4DCT) provides the ultimate quality information in terms of target 3D shape variation and 3D trajectory due to breathing motion. As with every technology, there are factors determining the level of success in this particular objective. Depending on the technology, even 4DCT might be compromised by residual motion artefacts and uncertainty associated with sorting images in a given phase bin, and so on. Moreover, even with 4DCT the question of representativeness of recorded breathing and associated motion patterns when compared with the conditions during the actual dose delivery is very relevant. For the purposes of this chapter, regardless of the quantity, quality and nature of source 3D images, we can assume that at some point there are a relatively low number of volumes corresponding to the various conditions expected during the actual dose delivery. These few volumes will be merged in order to ensure a given treatment isodose covers all expected positions of the target. These positions are in the coordinate system of the given dose delivery system. Whether this system is *static* or *dynamic* creates an interesting task which has been selected as another example of volume handling presented in this chapter.

Static Coordinates

Static coordinate systems have a fixed reference point around which the patient moves, for example, in a conventional linac isocenter. Provided there are 3D mask matrices for both (each) positions of a given volume with identical coordinate systems (registered images), then merging the two (all) volumes is a trivial logical sum ('logical *OR*') of the two (all) masks. A more interesting situation arises when a given volume motion occurs around the origin of a coordinate system which moves with that volume, for example, the case of target *tracking*.

Dynamic Coordinates – Tracking

Target tracking in radiotherapy means controlled radiation beam motion during beam on in synchrony with a moving target due to regular free breathing. This requires target position live monitoring, typically using internal or external surrogates. Common representatives of internal surrogates are metallic fiducial markers implanted directly in or in a close vicinity of the tumor. Fiducial markers are visible on an X-ray imaging system and their position can be watched live (provided the X-ray frame rate is sufficiently high and the additional radiation dose is justified). They can also be watched for periodic motion such as breathing, a series of X-ray images acquired and processed to create a dynamic model of target motion correlated with another, much faster (and possibly not based on ionizing radiation) system performing live monitoring. What we are interested in here is the fact that for these situations it is not the actual tumor itself that is actually being tracked by a given treatment beam, but rather a monitored internal surrogate – the fiducial markers. So, assuming a treatment plan is made using a single phase CT model only, the process works only under the assumption of a fixed bind between the tracked object (the surrogate) and the desired target or, to be even more accurate, a bind between the tracked object and *every part* of a given target. During treatment, the dose is being delivered to a volume relative to fiducial markers determined as a result of the treatment planning. Whenever the fixed bind is corrupt, there is a chance of a geometric miss of a moving target simply because this has not been accounted for during the treatment planning. Note that the approach addressing this aspect is referred to as *4D planning*, but probably for its relatively larger complexity and possibly also for its relatively low improvement of the overall accuracy of most cases this is not used widely in contemporary photon radiotherapy. For purposes of the example of volume handling, let's consider target tracking based on fiducials and a single phase CT model treatment planning. Then, in order to reduce the risk of missing a part of a target during the tracking of fiducial markers with a radiation beam, it is possible to consider a target volume relative to fiducial markers in another (breathing) phase than the one for treatment planning, or, dose optimization. The easiest way to address this is to acquire two CT series at complementary breathing phases (inspiration, expiration) with all fiducial markers visible, and define (clinical) target volumes (CTVs) for both CT models. One then can merge them the same way as in the above case of ITV and static coordinates system. The only difference being that in this case, we are merging two volumes in a dynamic system of coordinates (defined by fiducial markers) and that it is not the target motion itself that is the variable, but the variation of a given target position and shape relative to the tracked object, that is, the fiducials.

There are at least two ways to deal with this task:

1. Image registration with the fiducial markers as a matching objective followed by a trivial sum of the volumes ('logical *OR*') as in the static case, or

2. Determine each target voxel position in the fiducial-based coordinate system on the CT model *not used* for treatment planning (eg, CT at inspiration breath hold, *ibh*), and reproduce this position in the primary CT model (eg, CT at expiration breath hold, *ebh*) relative to the coordinate system determined by the primary set of fiducials. If the voxel found belongs to the primary target volume (CTV_{ebh}) then all is OK: this relative position is irradiated during both breathing phases. The opposite result means the position should be

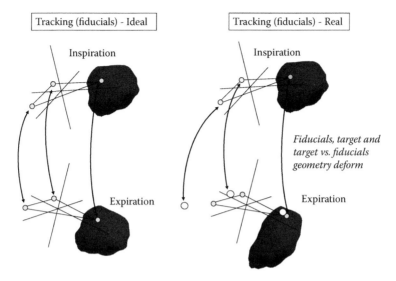

FIGURE 6.6 The relative position and orientation of the target with respect to (tracked) fiducials may vary during the breathing cycle. As the dose distribution is planned around fiducials on a single phase CT image, this geometrical deformation should be accounted for in the residual target margin. Another complicating factor is deformation of geometry of fiducials themselves.

added to the original target (CTV_{ebh}) to reduce the risk of missing that particular position during the second breathing phase (eg, inspiration). The principle is also shown in Figure 6.6.

The first approach is equivalent to registration based on paired reference points that can be found in the section 'Registration Based on Pairs of Reference Points' in Chapter 5. The second approach offers few options. One of them is assessing what combination of fiducial markers selected for tracking is associated with the smallest extra volume to be added to the original target volume in order to reduce the risk of it being missed during the opposite breathing phase.

Let's say we have the following problem:

- There is *N* number of fiducial markers implanted in a given target area

- The given area is affected by regular breathing motion and the fiducials will be used for given target tracking during dose delivery

- There are two CT series available: CT_{ibh} (acquired at inspiration breath hold – *IBH*), CT_{ebh} (acquired at expiration breath hold – *EBH*)

- The primary CT series, a 3D model for treatment planning, is the CT_{ebh}

- The primary and secondary CTs are registered by motion intact anatomy

There are two questions:

- What is the best combination of fiducials so that the target-fiducials bound deformation is minimal in the inspiration phase of breathing?

- What minimum anisotropic margin should be applied to expand the target volume in expiration to cover all parts of the target in inspiration?

Let's create a new script presenting an example solution.

Algorithm Scheme

Function Syntax
  targetXfiducials
Function Input/Output

  Prerequisite
    DICOM CT image series files and one DICOM RTSTRUCT file in one working folder. The following structures are expected contoured and contained in the RTSTRUCT file: target volume in EBH and IBH, fiducial markers in EBH and IBH, respectively
  Output
    bestCombi   A string variable indicating which combination of fiducials corresponds to the smallest geometry deformation, for example, '_1_3' for fiducials 1 and 3
    margin6   A 6-element vector [A, P, L, R, S, I] indicating minimal anisotropic margin to expand the target volume in EBH to cover the extra volume due to geometry deformation in IBH

Algorithm Steps

- Reconstruct all VOIs using the *RSdicom2mask* that is demonstrated in the section 'Volumes of Interest (RT Structures)' in Chapter 3
- Calculate the center of mass (CoM) for all VOIs (both target volumes and all fiducials in both breathing phases). For the reference 3D image matrix (*refImage*) and given *mask1D* ...

```
mask3D = zeros (size (refImage)); mask3D (mask1D) = 1;
[I, J, K] = find (mask3D == 1); CoM = mean ([I, J, K])
```

- Specify which VOIs (masks) belong to: *TVebh, TVibh, Fid_1_ebh, Fid_1_ibh, Fid_2_ebh, Fid_2_ibh, ... , Fid_N_ebh, Fid_N_ibh*
- Loop for all combinations of fiducials (eg, for three fiducials: 1, 2, 3, 1-2, 1-3, 2-3, 1-2-3)
  - For combinations with 1 or 2 fiducials consider only 3D translation ...
    - Calculate the *Origin_ibh* as the CoM of participating fiducial(s)
    - Calculate the *Origin_ebh* as the CoM of participating fiducial(s)

- Loop for all voxels of the *TVibh* (*voxel_TVibh*)
  - Calculate its 3D distance to the *Origin_ibh*

```
voxel_TVibh - Origin_ibh
```

- Reproduce the voxel's position relative to the *Origin_ebh*

```
round (Origin_ebh + voxel_TVibh - Origin_ibh)
```

- Check if the new voxel's position belongs to *TVebh* ...
  - If yes then do nothing, otherwise ...
  - ... record the voxel matrix indices as part of the *extraTVebh* representing target-fiducial(s) deformation for given combination
- For combinations with 3 or more fiducials, consider 6D translation & rotation as demonstrated in Figure 6.7
  - Calculate the *Origin_ibh* as the CoM of participating fiducials
  - Calculate a reference plane defined by 3 non-co-linear fiducials (include tests for collinearity). The reference plane is determined by a normal vector calculated as a cross product of two vectors defining the given plane

```
cross (Fid_2_ibh  - Fid_1_ibh, Fid_3_ibh  - Fid_1_ibh)
```

- Calculate the *Origin_ebh* as the CoM of participating fiducials
- Calculate alternative reference plane as for the *IBH* case above
- Loop for all voxels of the *TVibh* (*voxel_TVibh*)

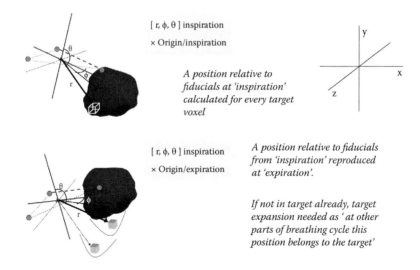

FIGURE 6.7   A scheme for a possible solution of the problem described in Figure 6.6 and the *targetXfiducials* algorithm example in the text.

- – Determine spherical coordinates $[r, \phi, \Theta]_{ibh}$ of the voxel in the system defined by the *Origin_ibh* and two directional vectors of the reference plane
- – Reproduce the voxel's position relative to the complementary system defined by the *Origin_ebh* and associated two directional vectors
- – Check if the new voxel's position belongs to *TVebh* …
  - – If yes then do nothing, otherwise …
    - – … record the voxel matrix indices as part of the *extraTVebh* representing target-fiducial(s) deformation for given combination
- Of all fiducials-combination specific *extraTVebh*, find the one with the smallest volume and display the associated combination as the first output: *bestCombi*
- Merge the 3D mask for the original *TVebh* with the 3D mask of the *extraTVebh*

```
mask3Dmerge = mask3D_TVebh | mask3D_extraTVebh
```

- Display the slices through the *mask3Dmerge* and the *mask3D_TVebh* in one figure by using the *imagesc* and/or 1-*contour* plot
- Find iteratively the appropriate anisotropic margin to expand the *mask3D_TVebh* to cover the *mask3D_extraTVebh* by using the *anisotropicMargin* and the *confParams* functions (see the section 'Expanding and Reducing ROIs' and Table 6.3). The anisotropic margin looked for (*margin6* output) is the one with the best compromise *COV* and *CI* parameters as defined above (see Figure 6.5)

The sum of *TVebh* + *extraTVebh* volumes can be interpreted as the ITV volume in the dynamic coordinate system defined by given combination of fiducials (subjected to available data). Naturally, such a volume is in principle smaller than the conventional ITV assumed in the static coordinate system. This is why target tracking (even though using fiducials surrogates) technologies have been developed for advanced radiotherapy systems.

Table 6.4 illustrates examples of the method applied on ten clinical target volumes treated among three CyberKnife centers. The table demonstrates significant normal tissue sparing when applying the optimum fiducials combination together with the optimum anisotropic margin in order to expand the target volume contoured at EBH phase to also cover associated parts of the target contoured at the complementary breathing phase (IBH) while tracking selected fiducials. Note two important aspects: tumor-fiducial(s) bound deformation during breathing (addressed here) is not the only uncertainty associated with the tracking, and, intra-observer variation in contouring is included in the method (possibly an advantage rather than the opposite).

The last thing to mention is that the principle of using fiducials as surrogates for a target volume position during treatment also applies to the sites not necessarily affected by the breathing motion. For example, fiducials are frequently applied in prostate radiotherapy. Whenever fiducials are used in this way, one should consider or assess to what extent they represent all parts of target at all possible conditions at treatment and whether the safety margins applied are adequate.

TABLE 6.4   Examples of 10 target volumes treated among three CyberKnife centers through fiducial tracking

GTV$_{ebh}$ [cm³]	Fiducial Combinations	extraTVebh [cm³]	Isotropic Margin to GTV$_{ebh}$			Anisotropic Margin to GTV$_{ebh}$		
			[voxels]	CI	COV	[voxels]	CI	COV
3.1	1, 2, 4	0.5	1	1.52	98.7	[1,0,1,0,0,1]	1.20	98.1
0.5	2	0.03	1	2.19	100.0	[1,0,1,0,1,0]	1.49	100.0
2.1	1, 2, 3	1.2	3	2.91	96.3	[3,1,3,0,0,3]	1.73	96.1
31.9	2, 3	16.4	5	2.58	92.6	[4,5,4,0,0,5]	1.77	90.7
137.6	3	17.9	3	1.61	99.8	[3,1,2,1,0,2]	1.24	99.3
81	1, 2, 4	64.5	2	1.44	99.6	[2,2,0,2,0,0]	1.21	99.2
105.6	2, 3, 4	23.0	5	2.02	98.7	[4,3,0,0,1,5]	1.30	98.1
75.9	1	9.0	3	1.79	99.2	[1,3,3,1,2,3]	1.52	98.5
55.7	1, 2, 4	25.9	4	1.99	94.3	[3,4,3,4,3,0]	1.82	93.5
95.6	1, 2, 3	14.9	3	1.71	99.1	[2,3,3,1,3,0]	1.57	99.0

Source: Dvorak P. et al., Intelligent margins for SBRT based on tracking fiducials, Paper presented at the SRS/SBRT Scientific Meeting, Radiosurgery Society, Minneapolis, MN (2014).

Note: The second column shows the best combination of available fiducials corresponding to the minimum *extraTVebh*. The next columns show the difference in *CI/COV* parameters for target volume (planned at EBH phase) expansion by an isotropic and anisotropic margin calculated by the method presented in this chapter, in order to achieve comparable coverage of targets merge contoured in the EBH and IBH breathing phase. Note significant normal tissue sparing by applying appropriate anisotropic margin when compensating tumor-fiducial deformation between two extreme breathing phases.

REFERENCES

Dvorak P, Richmond A, Gaya A, Santaolalla I, Floriano A, Knybel L. et al. (2014). Intelligent margins for SBRT based on tracking fiducials. Paper presented at the SRS/SBRT Scientific Meeting. Radiosurgery Society, Minneapolis, MN.

ICRU Report 62. (2010). Prescribing, recording, and reporting photon beam therapy. *Journal of the ICRU* 10(1).

Ohtakara K, Hayashi S, Hoshi H. (2012). The relation between various conformity indices and the influence if the target coverage difference in prescription isodose surface on these values in intracranial stereotactic radiosurgery. *British Journal of Radiology* 85(1014), e223–e228.

CHAPTER 7

Three-Dimensional Dose Calculation in Radiotherapy

LEARNING OUTCOMES

LO 7.1   Explain the role of dose calculation in radiotherapy
LO 7.2   Explain voxel-based and factor-based dose calculation algorithms
LO 7.3   Explain CyberKnife fixed collimator beam data set: OCRs, TPRs and DM
LO 7.4   Explain CyberKnife fixed collimator treatment plan parameters
LO 7.5   Prepare interpolated beam data lookup tables: OCRs, TPRs and DM
LO 7.6   Explain equivalent path length inhomogeneity correction
LO 7.7   Calculate voxel dose beam contribution for given geometry and beam parameters
LO 7.8   Calculate the dose distribution on a CT model for a given treatment plan including volumes of interest
LO 7.9   Explain the general aspects of independent dose/MU check in radiotherapy treatment planning
LO 7.10 Perform a simple calculation for fast routine independent dose/MU check in clinical practice

INTRODUCTION

This chapter presents an example of a three-dimensional voxel-based dose calculation algorithm in MATLAB. This particular implementation was inspired by the RayTracing® algorithm implemented clinically in the commercial treatment planning system which is used to model CyberKnife stereotactic beams shaped with circular collimators of various sizes. The algorithm includes a simple 1D inhomogeneity correction based on equivalent path length. The major motivation for selecting this particular system as an example was its simplicity. The dose calculation is based on beam parameters looked up directly from measured data tables. Options for using the application as an independent dose/MU check and relevant aspects to consider are discussed at the end of the chapter.

A BRIEF INTRODUCTION TO DOSE CALCULATION

The first requirement of radiotherapy is the availability of a source of suitable radiation. Knowing and deciding where, when and how large a radiation dose to deliver is the second requirement. The third requirement is to know how to do it using the given radiation source. How many radiation beams, how large and what shape the radiation fields, beam fluence modulation (yes, no, what type?) and how much radiation from a given beam in terms of radiation beam-on time or, for modern linear accelerators, monitor units (MU)? This third requirement is the process known as radiation treatment planning. To know how a given treatment machine and choice of exposure parameters translate in a particular absorbed dose at a given point in the patient body is dose calculation. It is the core of the treatment planning process.

Voxel-Based Dose Calculation

A patient CT model provides essential quality information not only for knowing and deciding *where to treat*, but also for providing a platform for three-dimensional dose calculation. Introducing computers in radiotherapy planning enabled calculating digitally reconstructed radiographs (DRR) from the CT data, that is, patient 2D projection 'images' from any direction, which together with displaying overlaid VOIs projected contours and beam shaping devices provides the base for calculating the dose distribution in 3D. Using a patient's 3D model for multiple voxel dose calculation by a computerized algorithm is then only a natural extension leading to the possibility of qualitative (isodoses display) and quantitative (DVH) plan quality assessments. Having available a 3D distribution of CT numbers leading to tissue-specific beam attenuation factors in high resolution also enables inhomogeneity corrections to account for water non-equivalence of various human tissues, which again dramatically improve the accuracy of dose calculation. Although many (but not all) modern TPS allow dose calculation grid size and resolution specified by the user, for purposes of this chapter, we assume dose calculation point equivalence with patient CT model voxels. Calculating dose for a group of selected CT model voxels eliminates registration issues. It is intuitive and generally does not compromise resolution as CT voxel dimensions are typically very small, around $1 \times 1 \times 1$–3 mm^3.

Factor-Based Algorithms

Although modern dose calculation algorithms are much more complex and based on sophisticated modeling of particle transport where usually at least some part involves Monte Carlo methods, non-modulated fluence beams can be modeled with usually acceptable accuracy even using a simple factor-based approach. Factor-based algorithms are based on the simple principle of using a specific factor to correct for the deviation of particular beam parameters during the exposure from the beam reference, that is, calibration conditions. Calibration conditions consist of a set of radiation beam parameters that are either good from the dosimetry perspective (expected accurate), or they represent typical clinical conditions, or, ideally, both. Correction factors should be available for each beam parameter value with an effect on dose. Typical parameters include distance from the source, depth in tissue, radiation field size and shape and, of course, indicator of amount of radiation, beam-on time for radionuclide sources, or monitor units (MU) for accelerators. Additional correction factors might be required for an additional accessory such as hard wedges to modulate fluence in one dimension. There can also be factors associated with patient anatomy to correct for difference between calibration environment which is often exclusively water and actual patient tissues.

However, some beam parameters associated with modern technology, such as IMRT, are so complex or consist of so many degrees of freedom that it is no longer possible to follow the factor-based approach. Dose calculation must then be performed in a different and more sophisticated way efficiently handling a large amount of beam parameters – beam modeling using computers. In this chapter, we will demonstrate a computer implementation of a simple factor-based three-dimensional, voxel-based, dose calculation algorithm.

EXAMPLE OF 3D DOSE CALCULATION

The RayTracing algorithm, which is a factor-based algorithm for the CyberKnife system, addresses the specifics of the radiation beams delivered by the treatment device. CyberKnife, in its basic version, produces multiple flattening-filter-free (FFF) photon beams of nominal energy at 6 MV. Individual fields are shaped by circular collimators producing cone beams of a given set of diameters at reference 800 mm *SAD* (Source-Axis-Distance) distance from the source: 5, 7.5, 10, 12.5, 15, 20, 25, 30, 35, 40, 50 and 60 mm. Unlike conventional clinical accelerators, CyberKnife is a non-isocentric machine so the axes of a given set of radiation beams generally do *not* cross at a single point.

The general dose calculation formula is as follows (slightly modified from CyberKnife PEG):

$$D = MU \cdot OCR(coll, R_{800}, d_{eff}) \cdot \left(\frac{800}{SAD} \right)^2 \cdot TPR(fieldsize, d_{eff}) \cdot DM(coll, SAD)$$

where the individual elements are:

D ... absorbed dose in Gy at given calculation point (voxel)

MU ... the given beam monitor units

d_{eff} ... effective, 'water equivalent' depth of a given calculation point (voxel) in tissue measured along a given rayline connecting the calculation point with the source (hence '*ray tracing*')

OCR ... a given collimator specific *Off Center Ratio* for a given radius (distance from beam axis) at reference *Source Axis Distance* SAD_{ref} of 800 mm (R_{800}) and given effective or 'water equivalent' depth in tissue d_{eff}

SAD ... actual *Source Axis Distance*; a distance from the source of a given calculation point projection along the beam axis, or, a distance of the source from the plane perpendicular to the beam axis containing the given calculation point

TPR ... *Tissue Phantom Ratio* for a given field size at actual *SAD* and given effective depth in tissue d_{eff}

fieldsize ... actual field size at actual *SAD* at plane perpendicular to beam axis containing the given dose calculation point *coll* … a given collimator identifier – numeric equivalent to field diameter at reference SAD_{ref} (800 mm)

DM … field output factor for a given collimator and actual *SAD*

A general introduction to basic radiation treatment beam parameters and factor-based formalism for dose calculation can be found in many teaching materials, for example, in IAEA 2005.

The RayTracing algorithm presented here contains only a basic inhomogeneity correction based on equivalent path length (EPL) calculated as 'water equivalent' or effective depth of a given calculation point along the rayline connecting a given beam source position with the dose calculation point.

$$EPL = (d_{eff}) = \frac{\sum_i \rho_i^{el} \cdot x_i}{\sum_i \rho_i^{el}}$$

Where

i … represents each CT model voxel along a given rayline

ρ_i^{el} … relative electron density of the *i*-th voxel

x_i … voxel dimension along a given rayline

Standard CyberKnife dose calibration is such that 1 MU delivers the dose of 1 cGy at the following reference conditions: 60 mm collimator, beam axis, water at the depth of 1.5 cm and reference SAD (SAD_{ref} = 800 mm). See Figure 7.1 for a graphical presentation of the reference and general conditions with respect to dose calculation parameters from the above equation.

A general problem of implementing a simple computerized 3D dose calculation algorithm using the above formalism can be structured as follows:

Preparations Step

- Gathering beam data and organizing it in efficient lookup tables

- Creating or reconstructing patient CT model including RT structure masks

    - This is to address individual voxels for calculating relevant dose distribution and also for surface curvature and inhomogeneity corrections

- Gathering and organizing individual beam parameters from a given RT plan data

    - The beam parameters must reflect the CT model coordination system

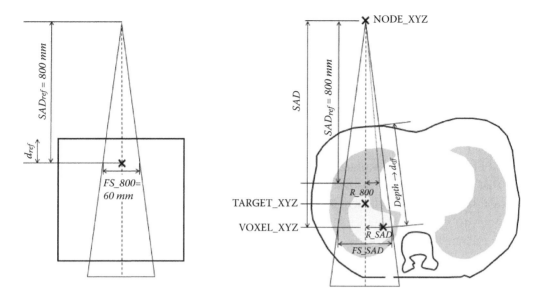

FIGURE 7.1 CyberKnife standard calibration conditions – water phantom, SAD_{ref} = 800 mm, 60 mm collimator (FS_800), reference depth d_{ref} = 15 mm (left) vs. general conditions (right). General conditions include non-reference SAD, collimator, off-axis distance, depth, MU, generally non-water material composition and non-flat surface.

Dose Calculation Step

- Addressing individual voxels and calculating dose for all radiation beams forming the given plan

- Interpolating, if needed, DVH calculation, dose distribution display, etc.

Considering what we have learned so far, the relatively more complicated are the two new processes from the preparations phase: organizing beam data and organizing plan parameters. In the following two sections we will describe possible simple approaches.

Organizing Beam Data in Lookup Tables

The algorithm example presented in this chapter requires a standard beam data set measured in a standard water phantom. For purposes of alternative dose calculation or a given plan dose/MU check, the beam data may be different from the beam data set used for the commissioning of the TPS. Of course, there is the aspect of independence, but even if the data set has an entirely different origin, it can certainly be organized in the same format as the original data for commissioning of the original algorithm. Of course, if independence is important, then only data unprocessed by the given TPS should be used. However, it does not mean that only raw data scans should be applied. There is always some processing required such as smoothing, but this should be done independent of the TPS. Also, if we are reproducing a given algorithm (as in this case), then it is logical to use the same structure and format of the data as the original implementation. Nevertheless, for situations where this particular algorithm reproduction will be compared with another dose calculation for the same treatment machine, then the level of independence is increasing as different algorithm implementation requires generally different beam data sets. In this training example we are reproducing a commercial algorithm implemented in a clinical system so that the original beam data set is used.

Off-Center Ratio (OCR)

Small FFF fields and the technique of superposing many beams in a target area is a combination that makes the OCR factor very important with a big impact on the overall accuracy of the calculated dose, so implementation decisions must be precise. Dose rate changes rapidly with increasing off-axis distance so spatial resolution of the OCR data is a crucial factor.

For each of 12 collimators, radiation beam profiles at five depths (15 mm, 50 mm, 100 mm, 200 mm and 300 mm) are required. Standard scan spatial resolution is 0.2 mm. From each dose profile, the OCR values are obtained as ratios of the CAX and average of two signals at given distance from the CAX. An example of the OCR measured data in the format required by the TPS is shown in Table 7.1.

It is obvious that the original measured data coverage is insufficient and interpolations are necessary in order to provide adequate OCR value for a given dose calculation point – radiation beam pair. After importing OCR tables in MATLAB comes the first important decision to make. Should a given lookup table be rather small and sparse leaving all interpolation to the main program, for a given radius and depth, lookup 2 and 2 neighbouring OCR values from the collimator specific table and perform 2D interpolation to obtain final OCR? Or is it better to interpolate the OCR table with sufficient resolution in both dimensions to make it ready for a *close-enough* direct lookup without the need for further interpolation in the main program? The choice depends on what is computationally more demanding, whether occupying a larger volume of computer memory or performing a given mathematical operation. In our case, considering the level of programming and basic computer equipment, the first choice wins. Table 7.2 is an example table (MATLAB 2D-matrix) ready for direct lookup of the OCR value for the given combination of radius and depth. Whole set of data is available in Web Supplement 7.1.

TABLE 7.1   Example of OCR table for the 10 mm collimator – OCRs are given in the range from 0 to 30 mm from the axis based on dose profiles at 5 different depths in a water phantom

	15	50	100	200	300
Version = 100					
Sample = 0					
10 Collimator Size (mm)					
5 Depths (mm)					
0	1.000	1.000	1.000	1.000	1.000
0.2	1.001	0.999	1.000	0.998	1.004
0.4	1.001	0.997	1.001	0.997	1.003
⏐	⏐	⏐	⏐	⏐	⏐
29.8	0.001	0.002	0.002	0.002	0.001
30	0.001	0.002	0.002	0.002	0.000

TABLE 7.2   OCR lookup table for the 10 mm collimator in MATLAB

9999	5	6	7	8	–	399	400
0.0	1.0000	1.0000	1.0000	1.0000	–	1.0000	1.0000
0.1	1.0001	1.0007	1.0007	1.0007	–	1.0049	1.0050
0.2	1.0015	1.0015	1.0014	1.0014	–	1.0099	1.0100
0.3	1.0018	1.0017	1.0016	1.0016	–	1.0094	1.0095
|	|	|	|	|	|	|	|
59.9	0.0000	0.0000	0.0000	0.0000	–	0.0000	0.0000
60.0	0.0000	0.0000	0.0000	0.0000	–	0.0000	0.0000

Note: Data was obtained by 2D interpolation in radius (range 0–60 mm, spacing 0.1 mm) and depth (range 5–400 mm, spacing 1 mm) from the measured and averaged data in Table 7.1. The 9999 value is arbitrary just to fill the rectangular matrix with the first column and row indicating radius and depth, respectively (*iOCRtables/iOCRtable* variable for the algorithm example below).

The second key decision is about direct lookup data coverage. Provided the basic measured data cover, for example, depths between 15 and 300 mm, the question of what happens with calculation points at depths outside this range is crucial. In general, there are few options: error (dose is not calculated), zero (dose is zero), extrapolation, last/edge value (nearest neighbour). Naturally, we do not want to exclude points either in the buildup region or beyond 300 mm, so simply we may have to accept larger uncertainty as a result of the simplistic physics and compromises applied. In the implementation presented here, the interpolated range of depth is 5–400 mm and the interpolated range of radii 0–60 mm. These ranges cover most clinical situations considering also beam divergence.

Tissue Phantom Ratio (TPR)

In the given formalism the TPR gives the dose in depth in water for a given field size. The standard approach is to recalculate TPR values from measured percentage depth dose (PDD) curves using known conversion formulas. An alternative solution is to measure TPR values directly using dedicated automated water phantoms at specified water level (IAEA 2010). An example of measured TPR values for all collimators and standard range of depths is presented in Table 7.3.

TABLE 7.3   Example of a TPR table

Version = 100					
Sample = 0					
12	Field Sizes (mm)				
301	Depths (mm)				
	5	7.5	–	50	60
0	0.457	0.414	–	0.380	0.406
1	0.594	0.538	–	0.484	0.501
2	0.720	0.655	–	0.582	0.592
|	|	|	|	|	|
299	0.235	0.248	–	0.299	0.306
300	0.234	0.247	–	0.297	0.304

Note: TPRs are given for depths ranging from 0 to 300 mm and each of the 12 fixed collimator sizes.

TABLE 7.4   TPR lookup table: Interpolated and extrapolated Table 7.3 for ranges of depths and field sizes likely covering nearly all possible clinical situations (see the *iTPRtable* variable for the algorithm example below)

9999	3	4	5	–	78	79	80
0.0	0.491	0.474	0.457	–	0.452	0.455	0.457
0.5	0.565	0.545	0.525	–	0.492	0.494	0.496
1.0	0.638	0.616	0.594	–	0.531	0.533	0.535
1.5	0.705	0.681	0.657	–	0.570	0.572	0.573
|	|	|	|	|	|	|	|
349.5	0.224	0.229	0.234	–	0.317	0.318	0.319
350.0	0.224	0.229	0.234	–	0.317	0.318	0.319

In a clinical situation, the TPR value required for lookup is for the actual field size at the plane of dose calculation point and for given depth. TPR values vary with depth so in order to avoid interpolation within the program, the lookup table should be interpolated so that the error associated with data resolution and parameter rounding is clinically acceptable. Of course, due to the curve steepness there will always be higher uncertainty in the buildup region unless there is a variable grid. In terms of field size, it is important to realize that there is beam divergence in play. So, in order to keep options for dose calculation relatively far from reference SAD (ie, 800 mm from the source), which can be perfectly reasonable calculation point location, it is necessary to allow field size range significantly larger than the nominal size of all collimators. In this particular implementation, the field size range was chosen to cover 3 to 80 mm. An interpolated data lookup table is shown in Table 7.4.

Output Factors (DM)

Dependence of dose on collimator opening (eg, field size), is determined by 'field output factors'. In this particular formalism denoted as DM (CyberKnife PEG), the output factors are measured at three SAD distances. Table 7.5 shows measured values for the fixed collimator set.

In agreement with the dose calculation formalism, the DM factors are specific to nominal collimator size and distance from the source SAD. Therefore, only interpolation in SAD of the measured data is required as shown in Table 7.6.

TABLE 7.5   Example of DM – Output factors tables as measured for three SADs in the format required by the TPS

Version = 100					
Sample = 0					
12	Field Size (mm)				
3	SAD (mm)				
	5	7.5	–	50	60
800	0.713	0.845	–	0.994	1.000
650	0.695	0.817	–	0.997	1.002
1000	0.720	0.869	–	0.991	0.997

TABLE 7.6   Interpolated and extrapolated DM lookup table for range of clinically applicable SADs (see the *iDMtable* variable for the algorithm example below)

9999	5	7.5	10	–	40	50	60
500	0.677	0.789	0.846	–	0.994	1.000	1.004
501	0.677	0.789	0.846	–	0.993	0.999	1.003
502	0.677	0.789	0.846	–	0.993	0.999	1.003
503	0.677	0.789	0.846	–	0.993	0.999	1.003
|	|	|	|	|	|	|	|
1499	0.737	0.928	0.956	–	0.981	0.983	0.989
1500	0.737	0.929	0.956	–	0.981	0.983	0.989

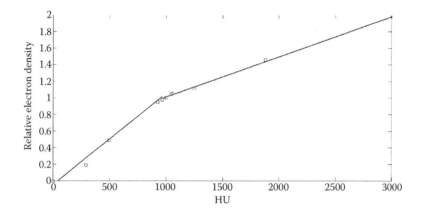

FIGURE 7.2   Electron density calibration curve used in the example algorithm implementation for simple inhomogeneity correction.

Equivalent Path Length Correction (d_{eff})

CT calibration for converting CT numbers or Hounsfield Units (HU) into relative electron (or physical) density is essential for dose inhomogeneity correction. Without it, the dose is calculated in water equivalent medium which is insufficient for the accuracy standards required by modern treatment planning. Standard HU to relative electron density calibration used in this example is shown in Figure 7.2. The two linear functions fitting the data in range of 0 to 1000 and 1000 to 3000 HU, respectively, were used in this algorithm implementation example.

Obtaining and Organizing Treatment Beam Parameters

Assuming that an important application of a computerized 3D dose calculation algorithm is treatment plan dose/MU independent verification, then the automated import of treatment beam parameters is an obvious requirement. This is particularly important when the beam parameter set is large. Treatment plan parameters are normally obtained from DICOM RT plan file (RTPLAN), see the section 'Importing DICOM Data/Other DICOM RT Data in Radiotherapy' in Chapter 3. However, DICOM RT does not necessarily support all treatment machines or simply the manufacturer does not support exporting plan parameters in this format. The example algorithm presented in this chapter is based on a non-isocentric CyberKnife with plan parameters which are not available in DICOM RT format. Instead, plan parameter

TABLE 7.7   Example of the CyberKnife treatment beam parameters for a clinical plan

| MU | COLL | NODE | | | TARGET | | | PID | NID |
		X	Y	Z	X	Y	Z		
371.06	15	352.42	−806.85	522.64	83.78	−54.58	125	1	1
170.15	10	−0.94	−885.05	432.18	82.51	−53.32	121.25	1	3
831.93	10	−0.94	−885.05	432.18	88.86	−63.47	115	1	3

Note: The first 3 beams of the total 64 are shown. Note that beams 2 and 3 originate from the same node but differ in direction. See the *PlanDataXYZ* variable for the algorithm example below.

data might be available in a XML format file in a given plan specific folder in a dedicated TPS computer and can be imported using MATLAB's inbuilt function *xmlread*. Sample plan files as well as sample plan data in MATLAB format are available in Web Supplement 7.2.

CyberKnife treatment beams are organized in treatment *paths*, each consisting of a larger number of *nodes*. A node represents the radiation source position from where the beam is directed towards the target.

Individual beam parameters for a circular collimator consist of:

- (beam specific) monitor units: MU
- (circular) collimator identifier (nominal diameter): COLL
- (beam specific) source (node) position: NODE_XYZ
- (beam specific) target reference point location: TARGET_XYZ

Note the two points determine both direction and distance of the source/beam to a given dose calculation point.

Two rather formal identifiers making navigation among beams easier are:

- Treatment beam path identifier: PID
- Treatment beam node identifier: NID

From the list of beam parameters above it is natural to organize plan parameters as presented in Table 7.7. The full plan data is available in Web Supplement 7.2.

Dose Calculation

Now we have all the necessary prerequisites ready for the actual point/voxel 3D dose calculation. From Chapter 3 we know how to reconstruct the patient model and RT structures from DICOM RT export in MATLAB: a 3D matrix of CT numbers and the same size 3D binary matrix for each VOI. We have beam data lookup tables with sufficient data coverage and resolution: a set of 2D-arrays with OCR data for each collimator, one 2D-array with TPR data and one 2D-array with DM (output factor) data. And we have the given treatment plan beam parameters organized in one table. The last thing we need is the CT calibration curve for the EPL correction, that is, a set of CT number vs. electron density pairs. In the next section, we will describe how to use all the available data to calculate the 3D dose distribution for a given treatment plan and a given CT model. Let's start with the algorithm for a function script in MATLAB calculating the dose for a given set of parameters for a given calculation point and a given treatment beam. An example of such a function based on the algorithm described above is available in Web Supplement 7.3.

Algorithm Scheme

Function Syntax

```
voxelBeamDoseCalc (MU, depth, SAD, R_SAD, coll, iOCRtables,
iTMRtable, iDMtable)
```

Function Input/Output

Input

MU Given beam monitor units

$depth$ Calculation point depth in tissue in millimeters. For inhomogeneity corrected dose calculation, this is the EPL

SAD Distance between the source and calculation point projected into the given beam axis (see Figure 7.1)

R_SAD Radius of the calculation point, that is, its distance in millimeters from the beam axis at plane perpendicular to it

$coll$ Collimator identifier; nominal diameter

$iOCRtables$ Set of interpolated OCR lookup tables (see Table 7.2)

$iTPRtable$ Interpolated TPR lookup table (see Table 7.4)

$iDMtable$ Interpolated DM table (see Table 7.6)

Output

A given beam dose contribution at the calculation point in Gy

Algorithm Steps

Preparations

- Calculate radius R_800 at reference $SADref$: $R\_800 = R\_SAD * 800/SAD$
- Calculate FS_SAD: $FS\_SAD = coll * SAD/800$

Data lookup indices

OCR
- Determine the *coll* (collimator) index for correct OCR data table from the *iOCRtables*

```
collSet = [5, 7.5, 10,...., 60];
% set of collimator sizes available in the same order as
% individual OCR tables in the iOCRtables variable
collIndexOCR = find (collSet == coll);
```
and select relevant *iOCRtable*
```
iOCRtable = iOCRtables {collIndexOCR};
```

- Determine the *R_800* radius index for correct OCR value lookup from the correct *iOCRtable*

```
radiusIndexOCR = interp1(iOCRtable (:, 1), 1:size (iOCRtable, 1),
                  R_800, 'nearest');
```

- Determine the *depth* index for correct OCR value lookup from the correct *iOCRtable*

```
depthIndexOCR = interp1(iOCRtable (1,:), 1:size (iOCRtable, 2),
                  depth, 'nearest');
```

TPR
- Determine the *depth* index for correct TPR value lookup from the *iTPRtable*

```
depthIndexTPR = find (iTPRtable (:, 1) == round (depth));
```

- Determine the field size (*FS_SAD*) index for correct TPR value lookup from the *iTPRtable*

```
fieldSizeIndexTPR = find (iTPRtable (1,:) == round (FS_SAD));
```

DM
- Determine the *coll* (collimator) index for correct DM value lookup from the *iDMtable*

```
collIndexDM = find (iDMtable (1,:) == coll);
```

- Determine the *SAD* index for correct DM value lookup from the *iDMtable*

```
SADindexDM = interp1(iDMtable (:, 1), 1:size (iDMtable, 1), SAD,
              'nearest');
```

Data values lookup

OCR
- Lookup OCR value …

```
if depth<=5
```

```
    OCR= iOCRtable (radiusIndexOCR, 2);
else
    OCR = iOCRtable (radiusIndexOCR, depthIndexOCR);
end
```

Note the minimum extrapolated data is used for points at depths below 5 mm. This is an arbitrary decision associated with some uncertainty since the shallowest depth of physical measurement was 15 mm. On other hand, this also allows calculating the dose at surface and OCRs vary with depth certainly much less dramatically than with off-axis distance.

TPR
  • Lookup TPR value …

```
TPR = iTPRtable (depthIndexTPR, fieldSizeIndexTPR);
```

DM
  • Lookup DM value …

```
 DM  = iDMtable (SADindex, collIndexDM);
```

Final dose calculation
```
 Dose = MU * OCR * (800/SAD) ^2 * TPR * DM/100;
```

Knowing how to calculate a single point (voxel) dose for a given treatment beam, we can proceed to putting together the algorithm to calculate multiple dose calculation points – 3D dose distribution. For clarity and a simple approach, spatial resolution but not for speed, let's put an equal sign between dose calculation *point* and *voxel*.

Algorithm Scheme

Prerequisites

Beam data:	*iOCRtables* (Table 7.2), *iTPRtable* (Table 7.4), *iDMtable* (Table 7.6)
Plan data:	Treatment plan parameters table (Table 7.7)
CT model:	Primarily for inhomogeneity correction, optionally for addressing calculation points (voxels) and dose display (see Chapter 3)
CT calibration data:	For simple inhomogeneity correction
RT structure masks:	Optional, for calculation points (voxels) addressing and DVH (see Chapter 3)

Algorithm Steps

Converting the CT matrix to matrix of relative electron densities
  • Recalculate *CTmodel* to relative electron densities *CTrelDen* using the CT calibration data (see Figure 7.2)

```
CTbelow1000 = find (CTmodel < 1000);
CTrelDen (CTbelow1000) = CTrelDen (CTbelow1000) * 0.001;
CTabove1000 = find (CTmodel >= 1000);
CTrelDen (CTabove1000) = 4.8667e-004 * CTrelDen (CTabove1000) + 0.5;
```

Selecting dose calculation voxels and their addressing
- Get matrix indices of the CT model voxels to calculate. for a given structure (VOI) mask (eg, *maskBody* indicating whole body volume within the CT model)

```
[rows, columns, slices] = ind2sub (size(CTmodel), find (maskBody
                == 1));
```

or just

```
[rows, columns, slices] = find (maskBody == 1);
voxels2calc = [rows, columns, slices];
```

Addressing calculation voxels is simple now; eg, matrix indices (*R*-ow, *C*-olumn, *S*-lice) of the 50th calculation voxel are obtained as: *VOXEL_RCS = voxels2calc (50, :)*

Calculating dose for each selected calculation voxel
- Loop for all selected voxels …
  - Calculate voxel XYZ coordinates in a given plan-CT model specific coordinate system (see the section 'Importing DICOM Data/Volumes of Interest [RT Structures]' in Chapter 3)

```
VOXEL_XYZ = [VOXEL_RCS (2)* PixelSize + CToriginXYZ(2),
        VOXEL_RCS (1) * PixelSize + CToriginXYZ(1),
        (VOXEL_RCS (3) - NoOfCTs) * z_spacing + CToriginXYZ(3)];
```

Note: Matrix indices (RCS) conversion to Cartesian (XYZ) and back is specific to IEC scale (IEC 61217) and patient orientation (*ImageOrientationPatient*) – see Chapter 3.

```
voxelDose = 0;% initial value
```

- Loop for all treatment beams …
  - From the plan parameters table (Table 7.7) get the beam specific source position *NODE_XYZ* and *TARGET_XYZ*
  - Calculate the calculation point off-axis distance *R_SAD* from analytic geometry (see Figure 7.1) …

$$R\_SAD = |NODE\_XYZ - VOXEL\_XYZ| \cdot sin\theta$$

$$\cos\theta = \frac{(NODE\_XYZ - TARGET\_XYZ)\cdot(NODE\_XYZ - VOXEL\_XYZ)}{\left|NODE\_XYZ - TARGET\_XYZ\right|\left|NODE\_XYZ - VOXEL\_XYZ\right|}$$

Then for the angle *theta* and expressed in MATLAB code …

```
theta = acos (dot ((NODE_XYZ - TARGET_XYZ), (NODE_XYZ - VOXEL_
        XYZ))/norm (NODE_XYZ - TARGET_XYZ)/norm (NODE_XYZ
        - VOXEL_XYZ));
R_SAD = norm (NODE_XYZ - VOXEL_XYZ) * sin (theta);
```

– Calculate SAD (see Figure 7.1) …

```
SAD = norm (NODE_XYZ - VOXEL_XYZ) * cos (theta);
```

– Calculate *depth* and d_{eff} (EPL) for inhomogeneity correction (see Figure 7.1) …
   At this point one needs to take a decision on whether to consider voxels along rayline only up to the body surface (contour) or whole way up to the source. The first option includes the testing condition for the body, the second one includes more voxels. For general aspects of the algorithm and the body contour smoothness aspects, here we decide to go without considering the body contour (mask).
   Calculate the *EPLdirVector,* a directional vector of a unit length (1 mm for standard CT scale) pointing from the calculation voxel towards the source. The optional *RTskipFactor* enables skipping voxels along the rayline to speed up density scaling (see Figure 7.3).

```
EPLdirVector = RTskipFactor * (NODE_XYZ - VOXEL_XYZ)/norm(NODE_XYZ
               - VOXEL_XYZ);
```

The increment determined by this vector will be used to trace voxels along the rayline, read their relative electron density and use it as a weight factor of that particular increment to add up to continuously increasing EPL (d_{eff})

```
EPL = 0;% initial value
EPLpreviousVoxelXYZ  = VOXEL_XYZ;% initial value
EPLnextVoxelXYZ  = VOXEL_XYZ;% initial value
EPLnextVoxelRCS = VOXEL_RCS;  % initial value
```

Check for the first *EPLnextVoxelRCS* falls within the CT model matrix and also FOV so it makes sense to calculate dose (see the full script in Web Supplement 7.3). If yes then proceed …

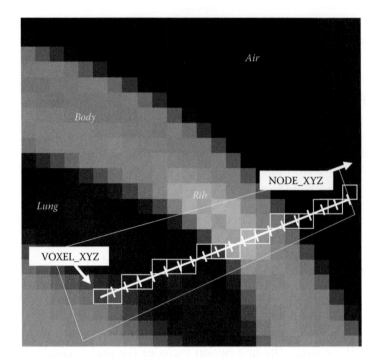

FIGURE 7.3   Selecting voxels along the rayline for contributing to EPL: The incremental directional vector between the point of dose calculation and the source can be of optional length in order to address approximately every i-th voxel along the rayline to read the associated relative electron density and use it as a weight factor for the given increment to the EPL (d_{eff}).

   – Loop: Starting from the *VOXEL_XYZ* calculate positions of the *EPLnextVoxelXYZ* by adding the increment (*EPLdirVector*) to the *EPLpreviousVoxelXYZ* ...

```
EPLnextVoxelXYZ = EPLpreviousVoxelXYZ + EPLdirVector;
```

... then calculate the corresponding CT matrix indices (R-ow, C-olumn, S-lice)

```
EPLnextVoxelRCS =
[round((EPLnextVoxelXYZ (2) - CToriginXYZ(2))/PixelSize),
round ((EPLnextVoxelXYZ (1) - CToriginXYZ(1))/PixelSize),
round (NoOfCTs - (1+(CToriginXYZ(3) - EPLnextVoxelXYZ (3))/z_
spacing)) + 1];
```

As long as the *EPLnextVoxelRCS* is within the FOV, that is, there is a CT number available at this position, add the relative electron density weighted increment to current value of the *EPL*

```
EPL = EPL + norm (EPLnextVoxelXYZ - EPLpreviousVoxelXYZ) *
     CTrelDen (EPLnextVoxelRCS);
```

Now, when the *EPL* has been increased by the *EPLdirVector* increment weighted by the *EPLnextVoxelRCS* associated relative electron density, we can update the *EPLpreviousVoxelXYZ* and carry on with the next step ...

```
EPLpreviousVoxelXYZ = EPLnextVoxelXYZ;
```

- Pick up the collimator nominal size (ID) from the plan data table (Table 7.7)

```
coll = PlanDataXYZ (beamNo, 2);
```

- Pick up the beam MUs from the plan data table (Table 7.7)

```
MU = PlanDataXYZ (beamNo, 2);
```

- At this point all parameters required for voxel beam dose calculation are read so the beam voxel dose can be calculated and added to total voxel dose ...

```
voxelBeamDose = voxelBeamDoseCalc (MU, EPL, SAD, R_SAD, coll,
                iOCRtables, iTMRtable, iDMtable);
voxelDose = voxelDose + voxelBeamDose;
```

- ... proceed with the next beam
- ... proceed with the next dose calculation voxel

At this point, we have calculated dose in pre-selected voxels, that is, the outcome is a 3D dose matrix with dose in selected voxels. By default, this dose matrix is registered with the CT model and RT structure masks in terms of DICOM RT reconstruction procedures described in Chapter 3. So calculating own DVHs, displaying multi-dimensional isodoses, and so on, can be done in the same way as with the original DICOM RT data reconstructed in MATLAB. Naturally, all these outcomes can be compared with the original equivalents exported from the TPS.

The two algorithms described above, that is, the function *voxelBeamDoseCalc* for calculating voxel beam dose and the general algorithm reproducing dose 3D calculation on background of a given patient CT model, demonstrate a simple implementation of a factor-based 3D dose calculation approach. This particular implementation, however, is simple and without optimization from an efficient programming perspective. So, although it produces results very similar to its commercial counterpart, the calculation time is much longer. Nevertheless, the main motivation of this chapter is to demonstrate a simple-to-follow example of a step-by-step process in order to reach the desired goal – calculating 3D dose distribution on background CT data, including a simple inhomogeneity correction.

Based on the application, even this demo dose calculation can be used in a way that enables calculation time reduction without compromising a given application objectives. Examples of applications or approaches follow:

- If a particular RT structure DVH is the objective, dose can be calculated only in the given VOI instead of arbitrarily defined large calculation grid or a whole body contour …

- Maybe dose needs to be available only in certain plane(s) instead of a large 3D volume …

- Like with the EPL calculation above, skip factors can be introduced and dose can be fully calculated only at every i-th voxel and interpolated using fast inbuilt functions at the end …

- Or, independently calculated dose may be required only at certain voxels as samples for verification purposes

The last approach has an interesting practical application and is described in the next section.

INDEPENDENT DOSE/MU CHECK

In modern radiotherapy, it is a common practice to perform an independent check of TPS dose calculation. This is even mandatory in some countries. This is usually done using third-party software where, typically, plan parameters (DICOM RTPLAN) are on the input, and beam-specific MU required to deliver associated dose contribution to the reference point are on the output side. Although recent versions have been rather simple and based on dose to flat water phantom and simple factor-based dose calculation, modern versions of such software include corrections for surface curvature and also inhomogeneity so further extra input data is required (DICOM RTSTRUCT, CT). The ultimate solution would be an independent fully 3D dose calculation of comparable class in terms of dose calculation accuracy as the primary dose calculation algorithm used in a given TPS. Considering this, an independent fully 3D dose calculation implementation in MATLAB platform as presented in this chapter can serve well for the purposes of dose/MU verification. Its major limitation (calculation speed), can be addressed and by-passed in a way so that the application objective is not necessarily compromised but possibly even improved.

Let's assume that we have an independent fully 3D dose calculation application for independent verification of dose/MU calculation by a given clinical TPS. Having two 3D dose distributions to compare, a major problem is how to compare them. Also considering that this is a task that would need to be done for every single RT plan produced, the application workload must be fast, smooth and robust.

In principle, there are two possible ways of looking at alternative dose calculations:

- Primary objective is *agreement* with the reference, ie, TPS, dose calculation, or

- Primary objective is absolute, ie, *clinical*. Alternative dose calculation is compared with primary clinical objectives to check whether these have been met or otherwise.

There are practical differences between the two approaches. The first approach has problems where there is a systematic difference in dose calculation algorithms. For example, comparing dose distributions in lung calculated with an algorithm belonging to Monte Carlo class, and doses calculated with simple 1D inhomogeneity correction, has generally an unpredictable outcome. In such situations, a user has few options:

- Based on known general differences in algorithm classes to estimate whether observed difference is reasonable

- Recalculation of the reference plan in a given TPS without inhomogeneities included, eg, with appropriate density override. Then it is more likely that potential differences observed are due to something else than the systematic difference of inhomogeneities handling by two algorithms involved and as such is worth special attention

Neither option is ideal but the second, clinical approach is relatively simpler in the sense that alternative dose calculation is skipped when selected clinical objectives are met, for example,

- Mean PTV dose is maximum, eg, 3% less than for the reference plan

- selected OAR D_{max} or V_{XGy} is maximum, eg, 3% larger than for the reference plan, etc.

Of course, a combination of both approaches is possible, for example, when the first one fails, the second one comes into place to guide the decision.

Another perspective is technicality of comparing dose distributions. Next to comparing selected point doses and selected DVH parameters, full 3D doses can be analyzed using, for example, gamma analysis comparison tool (see Chapter 9). In principle, whole DVHs can be compared by means of Tumor Control Probability (TCP) and Normal Tissue Complication Probability (NTCP) modeling, however, all these approaches are rather complicated for routine practice. Alternative dose/MU verification software have been traditionally based on reporting a given reference point beam dose contribution, mostly because it was the only approach that was offered. With 3D dose recalculation there are many more options but, consequently, also many more questions on how to report the result.

Returning back to the 3D algorithm example presented in this chapter, Figure 7.4 illustrates an example of a multiple lung lesion for a stereotactic radiotherapy case with both isodose and DVH (PTV and trachea) comparison between the original plan as calculated by the clinical TPS and calculated by using algorithms developed in this chapter. Differences are small mostly due to the same original beam data set and RayTracing algorithm. An important application aspect is including DVHs calculations using alternative (MATLAB) algorithm and original (TPS-exported) dose. By comparing all three modes, one can potentially estimate what parts of a difference is due to the difference in volume calculation and what due to the difference in dose.

As demonstrated, the algorithm implementation presented in this chapter can be in principle used as an independent 3D dose/MU check for the CyberKnife with fixed collimators on the basis of alternative full dose recalculation. However, in order to achieve the

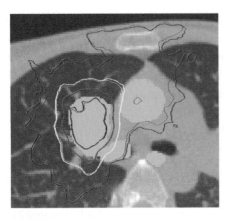 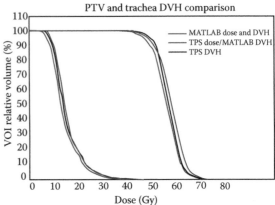

FIGURE 7.4 An example of a comparison between an original dose calculated using a clinical TPS and a dose calculated using the alternative algorithms implemented in MATLAB. Isodoses (left) in four colors show levels of 20%, 40%, 60% and 80% of D_{max}: TPS (thin lines), MATLAB (thick lines). Green areas show target and contoured OARs. DVH comparisons for PTV and the trachea as an OAR representative are shown on the right.

level of clinical routine practicality, it requires optimized programming to increase calculation speed significantly. It can be still applied, but one has to restrict the total number of calculated voxels. The following scheme has been found efficient in clinical practice:

- Only selected voxels are calculated using an independent 3D dose calculation algorithm

- Selected voxels are compared dose to dose with a general tolerance set to 5% difference of the reference dose

- Normalization dose, typically global D_{max}, voxel is always calculated. Another 6 voxels, always 5 voxel distances in each major direction, are included

- For selected VOIs a voxel at VOI's center of mass plus another 6 voxels (5 voxel distances in each major direction) are included

Such a clinical implementation example provides neither DVHs nor DVH statistics but for the reasons expressed above (questions how to interpret the result and possible washing out or averaging effects when using *macro* parameters), it is certainly a reasonable alternative. The most important fact is that this particular implementation is still fully three-dimensional and voxel-based which means substantial advantage over conventional 'beam-dose contribution to a reference point', especially for frequently occurring cases with multiple targets and single normalization (reference) point. The next advantage is that the level of independence is as high as it can be in spite of using identical beam data set. Dose calculation represented by this example is based on fundamental inputs, that is, CT model, DICOM RT structures, original plan parameter set, and CT electron density calibration curve. No parameters partially calculated by the TPS such as depth or off-axis distance are included.

REFERENCES

Accuray, CyberKnife, MultiPlan and RayTracing.

CyberKnife. Accuray Inc. http://cyberknife.com

IAEA. (2005). *Radiation oncology physics: A handbook for teachers and students*, Podgorsak EB (Ed), Printed by the IAEA in Austria, July 2005, ISBN 92-0-107304-6.

IAEA. (2010). *Radiation biology: A handbook for teachers and students*. IAEA-TCS-42.

IEC 61217:2011. (2011). Radiotherapy equipment – Coordinates, movement and scales. International standard by International Electrotechnical Commission. https://webstore.iec.ch.

RayTracing. CyberKnife: Physics Essentials Guide. Accuray, Inc. P/N 029577A-ENG.

CHAPTER 8

Semi-Automated Measurement of the Major Mechanical Parameters of Linacs

LEARNING OUTCOMES

LO 8.1 Explain the advantages of automation in radiotherapy equipment QC

LO 8.2 Explain the major mechanical parameters of a conventional isocentric linac

LO 8.3 Explain the conceptual differences between 'traditional' and image/based approaches

LO 8.4 Explain the motivation for the *one test–one method–one reference* approach

LO 8.5 Explain how to detect automatically a high-contrast object in a digital image

LO 8.6 Explain how to design a simple interactive interface to enter the reference points manually

LO 8.7 Explain how to measure distances and angles from the given test images

LO 8.8 Explain the importance of independence and order of tests for checking the validity of a QC program

INTRODUCTION

This chapter presents an *example* of a semi-automated quality control (QC) tests for the measurement of the major mechanical parameters of a conventional isocentric linac. At first, a general concept of automation is introduced followed by comparing a traditional and image-based approach. Emphasis is given to the choice of clinically relevant reference and order of tests in order to maximize independence and test validity. Next, *examples* of individual tests for *selected* major parameters are presented, each consisting of a method description, data acquisition and a dedicated computerized analysis using scripts in MATLAB.

SCOPE OF AUTOMATION IN RADIOTHERAPY QUALITY CONTROL

Automation of the processes in radiotherapy has similar advantages as in any other field: more efficient workflow, standardized quality, and saving of time and money. Quality control (QC), in terms of processes preventing errors in radiotherapy, includes many measuring procedures on a linac with a similar scheme:

- Phase 1: Phantom setup on a given machine

Note: Some procedures may not require a dedicated phantom.

- Phase 2: Acquire data (images, dose, …) using dedicated application or control software

- Phase 3: Analyze measured data using dedicated application or analysis software

- Phase 4: Save results and generate a report

There are few innate requirements for QC checks: they should be sufficiently accurate, efficient, and independent. Validity includes the optimum periodicity of the given tests so that the decision whether to regularly check a particular parameter, and if so how often, is crucial. If a given test requires a dedicated phantom then there is not much to optimize in phase 1 (above). Phase 2 already allows some optimization in terms of acquisitions for various conditions (imaging techniques to test at the time, including phantom offset or otherwise, etc.), data ranges, and so on. It is worth thinking when considering workflow aspects against the clinical ones, for example what is the clinical impact when a particular parameter goes wrong, what is the probability that this happens, and so on (AAPM TG100). Another option is introducing automation in the data acquisition phase. This is usually possible for tests defined by a manufacturer. On top of it, modern machines are already equipped with special modes allowing some automation eg, developer mode, research mode, and so on. These modes enable some access to the machine software to test the machine performance in sequences and combinations that are not permitted otherwise. Using such an approach one can consider defining a sequence of automated image acquisition for a series of gantry, collimator, and table geometries to test linac isocenter. Some level of automation of data acquisition provides opportunities to optimize phase 2 of the general test procedure above.

However, the major field for automation is phase 3: data analysis. As long as the data acquired is standardized, that is, there is only minor or predictable variation of the data contents, it is natural to introduce automation in this phase. Major advantages of such an approach is increased efficacy in terms of saving operator and machine time and, even more importantly, standardized quality. For example, acquiring a test image, displaying it on-screen of a given control software and using measuring tools to check distances or histogram data, represents a common approach used in many departments. Such an approach is usually associated with subjectivity. Automation allows defining clear objectives to extract relevant parameters and process them by using strictly defined objective analytical methods. However, 'objective' does not always mean 'correct' and robustness of the algorithm with respect to expected data variations that are not included in the test represents a major challenge. Another aspect is that acquired data may not be easily exportable from a particular acquisition system for third-party analysis.

Generating a report, that is, phase 4 and above, also offers some room for optimization. As a simple example, everyone probably can see the difference in efficiency between the following two example processes:

- Test parameter values displayed on-screen – retyped in a spreadsheet – printed in report – sign the report – archive the report

- Import the test data – run analyzing script – archive the report generated by the script (optionally signed electronically using logon data)

So, ideally report generating is integrated in data analysis, but this is not always straight-forward, as in when two types of analysis are required for one report. There are also standard tests which cannot be approached this way, for example, EMergency Off tests.

GENERAL ASPECTS OF LINAC QC

In the case of a new machine, there are few phases one usually goes through before creating a routine QC program for a conventional clinical linac. These are processes of:

- (machine) Installation

- (customer) Acceptance (procedure) testing

- Commissioning

- QC program setup

From a medical physicist perspective presented in this book, the installation procedure is entirely in the hands of the vendor. Naturally, there is support from the local team, but the subjected areas are rather in infrastructure and blueprints than performance characteristics of a given machine.

As soon as the machine is installed, the next phase is machine *acceptance* testing by representatives of the user. In the case of a radiotherapy accelerator, this is usually an experienced medical physicist. The acceptance procedure consists of demonstrating that major performance characteristics meet specification. Selecting parameters, methods and tolerances is entirely with the vendor. It is not uncommon that the tolerances are somewhat wider than the machine is actually capable of. However, what is important are the *methods* and *order of tests*. One should be careful to accept parameters on a one-by-one basis, that is, demonstrating and signing off when a particular parameter has an acceptable value. It is entirely in the hands of the installation engineer whether he keeps a previously accepted parameter unchanged while carrying out the installation and preparing for the acceptance of further parameters. Therefore, from a customer's perspective, it is natural to request that the full acceptance procedure be done in one session at the very end of the installation process when all parameters are ready for demonstration and sign off. As previously mentioned, the methods and measuring tools for the acceptance procedure are given by the vendor. This is okay with respect to the objectives of the procedure but might be a source of some uncertainty or inconsistency later.

Traditionally, as soon as a machine is accepted, the next phase called 'commissioning' starts. For a radiotherapy accelerator this is mostly about acquiring beam data for configuration of a given TPS, and acquiring baselines for a periodic QC program to be performed during the machine lifetime. As the methods and QC equipment used (eg, detector resolution as an example), influence the quality of the results, the choice of both should be done with care with respect to what is available and feasible within the department for the foreseeable future. This, of course, does not mean there cannot be changes anytime later, such as introducing new QC equipment or method, but in such a situation there is always the question of whether the original reference still applies or a new one has to be set. A good example is measuring radiation beam dose profiles using an automated water phantom with a small point detector for best spatial resolution as part of acquiring data for a given beam model. Measuring beam dose profiles, or parameters derived from them, is a part of standard recommended monthly QC tests (eg, AAPM TG 142) but it is inconvenient to use a large automated water phantom for such a frequent routine test. Instead, 2D detector arrays can be used efficiently for this purpose. Depending on the comparison analysis, it may not be straightforward to use the data from a water phantom as the reference for routine monthly testing using a detector array with generally different resolution. Therefore, at the time of measurement of the reference beam data for TPS commissioning, usually the reference baseline with a routine detector is acquired too. This is then considered the reference for periodic QC tests (eg, IPEM 2006).

However, one can probably detect a systematic small discrepancy between acceptance testing with method & equipment 1 and baseline measurement with method & equipment 2. Isn't it better and more straightforward to keep the methods, reference and tolerance constant through the whole process, ie, acceptance, commissioning and routine periodic QC? And is it possible? Why do we need baselines? We do not use them for acceptance. Why could we not use first-principles (ie, absolute, without comparison with baseline) references also for the routine QC program?

In general, the concept of baseline has been in use mainly because of significantly more demanding measurements for the acceptance or commissioning purpose compared with routine QC where time efficiency is an important factor.

In this chapter, we will demonstrate a system of *example* tests for the major mechanical parameters of a conventional C-arm radiotherapy linac. The basic mechanical parameters of a conventional linac were selected because of their simplicity and the fact that everyone working in radiotherapy would be familiar with these. It is also a part of every radiotherapy medical physics program and training scheme. Another reason is that in the opinion of the author, this particular area is convenient for demonstrating thinking and decision making regarding preferences of reference and methods of measurement with respect to the general aspects addressed above. In particular, this is important when any form of automation is to be introduced.

CONVENTIONAL MEDICAL LINEAR ACCELERATORS

The most common radiotherapy treatment machine today is a conventional isocentric C-arm linear accelerator producing high-energy photon beams. It consists of the isocentric gantry with a beam production line, beam collimation system (collimator), MV (mega-voltage) flat-panel detector mounted on a robotic arm to deploy and retract the detector, and the kV (kilo-voltage) source and flat-panel detector mounted on separated robotic arms. The MV detector (MVD) has been introduced to replace traditional portal films for pre-treatment verification of patient position relative to the beam isocenter. Today, MV detectors are still in use for this purpose but their application moves more towards beam dosimetry. From well-known physics principles, kV X-ray systems produce images with generally better image quality than MV systems. That is why modern radiotherapy machines are equipped with a kV source-detector pair (kVS, kVD) to produce either planar projection images or 3D cone beam CT (CBCT) for better quality image-guided radiation therapy (IGRT). A beam collimation system consists of three dynamic components which are typically included in QC programs: two pairs of secondary jaws (X and Y) defining a rectangular shaped radiation field, and a multileaf collimator (MLC) for fine beam shaping and/or beam fluence modulation. Integrated in the collimator is also the optical guidance system consisting of the light field source with the crosshair and optical distance indicator which, when projected on the patient surface, show the treatment field shape, beam axis, and source-surface distance (SSD) to guide setup. The treatment couch, or patient support, is another crucial component. A standard couch has 4 degrees of freedom: three translations (vertical, longitudinal, lateral) and isocentric rotation. There are also advanced (6D) models combining 3D translation with 3D rotation – tilt, pitch and roll. In addition rotation around the isocenter can be replaced or extended by rotation around other axes, but for purposes of this chapter the most common isocentric 4D couch will be considered as the standard. The last major component to be included to the QC program presented in this chapter is the laser aligning system. This is the system of three to four lasers placed on the treatment room walls indicating major anatomical planes through the machine isocenter. The two lateral lasers (left and right) indicate the transversal and coronal plane and they should be perfectly coincident. The sagittal plane is usually indicated by the sagittal laser placed on the wall facing the linac's gantry, and also by another laser placed on the ceiling to substitute the sagittal laser when covered by patient or accessory. All major components mentioned require precise mechanical setup and periodic QC measurements to ensure they perform within specs and agree with the machine model in a given TPS. The latter, that is, the machine model in the TPS, is actually the primary and gold standard reference. So, wherever possible, machine parameters and performance characteristics should be referred to this. Figure 8.1 shows a picture of a modern conventional radiotherapy accelerator indicating the major components just introduced.

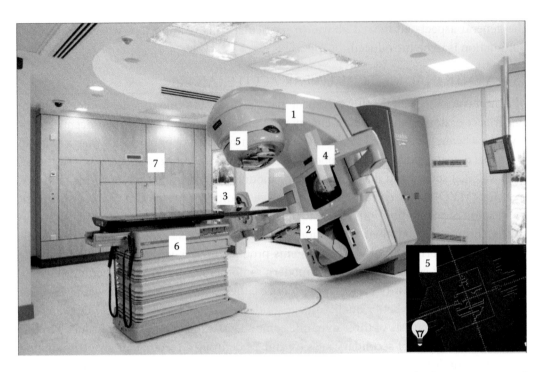

FIGURE 8.1 Conventional isocentric radiotherapy accelerator with 4D couch, dynamic MLC and kV imaging system. (1) Gantry, (2) MV detector arm, (3) kV source arm, (4) kV detector arm, (5) beam collimation system (secondary jaws X/Y and multileaf collimator) and optical field components, (6) conventional 4D couch, (7) alignment laser system.

MEASURING THE MAJOR MECHANICAL PARAMETERS

To ensure that the machine's mechanical performance is within specs or, more importantly, in agreement with the machine model in the TPS, the following critical parameters must be verified:

- Gantry rotation scale to ensure the beam radiation axis points towards the isocenter exactly under the angle indicated on the machine digital scale

- Collimator rotation scale to ensure the beam collimation system angle is exactly as on the machine digital scale

- Secondary jaws (X/Y) scale to ensure the individual jaw position relative to the central beam axis (CAX) is exactly as indicated on the machine digital scale

- Collimator isocenter to ensure that the beam CAX walkout for full range of collimator rotation angles is as small as possible and within tolerance

- MV (treatment) isocenter to ensure that the beam CAX walkout for full range of gantry rotation angles is as small as possible and within tolerance

- kV isocenter to ensure that the kV imaging system axis points exactly at the MV treatment beam isocenter for the whole range of gantry rotation so that any discrepancy in a kV image represents the discrepancy from the treatment beam CAX

- kVS and kVD arm positioning scale to ensure that both robotic arms set the source and the detector position correct relative to MV isocenter

- MVD arm positioning scale to ensure that the detector positioning relative to MV isocenter is as indicated on the machine digital scale

- Aligning lasers directions and rotations to ensure that all laser planes intersect at MV isocenter and represent accurately major anatomical planes

- Couch translation and rotation scale to ensure that the couch positioning relative to the MV isocenter is as indicated on the machine digital scale

- Couch isocenter to ensure that the couch isocentric walkout is as small as possible and within tolerance

- MLC leaves positioning in static and dynamic mode to ensure leaves calibration relative to the beam CAX is accurate for whole range of leaves positions and that their dynamic performance is correct

Now, knowing what component parameters are required to verify, we need to decide *how* and in what *order* so the result is the most representative for desired clinical applications. The test order is important mainly for the acceptance procedure. Routine periodic QC is less sensitive to this because the probability that two independent parameters move exactly by amounts that compensate each other is negligible.

THE TRADITIONAL APPROACH IN BRIEF EXAMPLES

In general, what is meant here by *traditional approach* can be briefly introduced by using a few aspects of the procedures applied during installation, acceptance, and/or periodic QC. For example, secondary jaws X and Y positioning calibration and checks have been traditionally based on the light field, cross-hair and graph paper, or, the imager positioning is based on the measurement of a physical distance of the panel chassis to a reference structure in the collimation system. The first approach introduces a small systematic error due to the different attenuation of light and MV X-rays through the jaws and also due to cross-hair reference, and is subjective in principle. The latter might do the same because although the distance between the chassis surface and the detector layer is considered for correction, it may not be the image itself that is used for calibration. Another example: the gantry rotation scale is usually checked using a spirit level attached to a reference surface on the machine collimation system. A principal imperfection of such an approach is that it is not the beam direction which is measured. So, however small the difference between the nominal beam direction and the given reference surface, another (small) systematic error is the result. Two more examples: so called 'spoke test' is used to measure gantry and/or collimator rotation axis walkout. It is based on a narrow field exposed on a film (or any other detector, irrelevant here) for the gantry or collimator angles sampling whole range of possible rotation. The problem is that such a test is necessarily influenced by the accuracy of the beam collimation device used. In principle there are few options to collimate the beam: secondary jaws X or Y, MLC, or maybe, a special dedicated collimator which can be used during installation only. The key point here is that the test result includes not only the gantry or collimator walkout, but also the potential asymmetry in beam collimation device. Therefore, the choice and accuracy of beam collimation, which also determines beam radiation axis, is crucial. Additional details are provided later in this chapter.

These are brief examples behind the motivation for a semi-automated image based system of checks together with optimized choice of reference and test order. An example of such a program for a conventional isocentric radiotherapy accelerator is presented in the next section.

IMAGE-BASED APPROACH

The general advantages of a QC program for mechanical parameters of a radiotherapy accelerator based on radiation images using both treatment (MV) and verification (kV) imaging systems are obvious:

- Parameters are measured using radiation beams directly, there are none or much fewer substitutes or surrogates when compared with the traditional approach

- Parameters can be quantified from dedicated test images using automated algorithms providing improved accuracy, robustness and objectiveness

- Test instrumentation and methods can be based on standard accessories available with the machine so minimum extra equipment is required

- The program washes out potential minor but systematic differences among the various phases of the introduction of the machine into the service: acceptance, commissioning, and routine periodic QC using *one test–one method–one reference* approach

Reference Systems

The example QC program presented in this chapter is based on the definition of one primary reference:

- Treatment (MV) radiation beam axis and reference directions are defined by the sides of the MLC symmetric leaves

The reason for this choice is that the MLC is used for fine beam collimation and for fluence modulation so, of all beam collimation devices have the highest clinical relevance (considering modern radiotherapy). It is also arguably a very precise component mechanically. The decision to prefer leaf sides to leaf ends is the independence of calibration. The only parameter determining the accuracy of such a primary reference is the MLC lateral placement relative to the collimator mechanical rotation axis, which can be verified with high accuracy using a dedicated spoke test (see the section 'Collimator Rotation Spoke Test' in this chapter). The *tongue-and-groove* effect included can also be quantified and accounted for if not found to be negligible.

Using the primary reference it is already possible to define the machine isocenter and also setup alignment lasers which can then serve as the *secondary reference* for some tests.

Test Images Import and Common Principles

The tests examples presented in this chapter are based on the automated or semi-automated detection of one of the following objects in a given test image:

- Radiation field border defined either by the MLC leaves (side or ends) or secondary jaws (X or Y)

- Center of a high-contrast sphere (ball bearing) of the only phantom required for the test examples presented

The tests are based on imaging of specific MLC test patterns alone, or in combination with imaging the ball bearing cube phantom where the ball bearing provides a test specific reference. A simple ball bearing cube phantom is a cube made of material with much smaller X-ray attenuation then a small metallic ball bearing placed in its center. Each of the six walls has visible markers indicating the ball bearing projected position on that given wall. Before using such a phantom, it is essential to verify the agreement of external markers with the actual position of the ball bearing as shown in Figure 8.2.

All test examples apply one or more of the following four basic procedures:

- Import DICOM test images

- Detect a beam collimation device edge

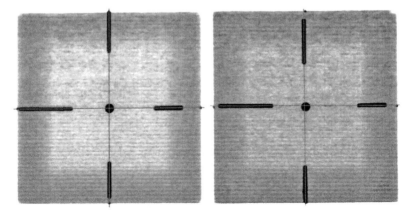

FIGURE 8.2   Examples of the ball bearing cube phantom test using X-ray imaging. Thin wires were placed on the phantom's external markers and X-ray images acquired. A good (left) and poor (right) test result is demonstrated for two example sides (directions).

- Detect a ball bearing center

- Manual input of reference pixel indices

The procedures are used in test-specific MATLAB scripts available in Web Supplement 8.1.

Import Test DICOM Images

- Import, rescale and convert test images in MATLAB 2D array

```
testImgInfo = dicominfo (testImgFilename);
testImg = double (dicomread (testImgFilename)) * testImgInfo.
        RescaleSlope + testImgInfo.RescaleIntercept;
```

Detect a Beam Collimation Device Edge

Detecting a field edge in an image is a basic requirement for measurement of the distance. A simple script calculating the FWHM from a radiation field profile was demonstrated in the section 'A Simple Case Study' in Chapter 1. The same principle applies here. The first step is to extract the relevant imager signal profile, the second step is to determine the reference signal (100%), the third step is to find pixels with the signal above a specific threshold (typically 50%, but may vary when only symmetry is an objective). The last and the first of the selected pixels then represent the given field edges searched for. When a more complex pattern with multiple field borders is present in a given profile (eg, MLC tests), a script has to include additional selection from the positive pixel coordinates above the signal threshold.

For objectives of most of the test examples presented here, the reference image signal is defined around the maximum. For better robustness let us define the reference intensity as the 100th percentile from the image histogram of 100 bins. Effectively, this is the maximum signal but not based on a single pixel value.

```
[counts, binCentres] = hist (testImg (:), 100);
refSignal = binCentres (100);
```

When a test specific *testProfile* is extracted, the in-field segment of the profile is found using

```
fieldSegment = find (testProfile > 0.5 * refSignal);
```

with the corresponding edge pixel indices *fieldSegment (1)* and *fieldSegment (end)*, respectively. The corresponding distance between the two pixel indices at the level of the imager is obtained by multiplying by the nominal pixel size. For better resolution, the profile can be interpolated (eg, factor 4 or 8). The interpolation factor must be then included for the distance calculation. When the desired distance is to be measured at a different level (distance to the source) than the imager at the isocenter with the detector 50 cm below it, then an appropriate scale factor must be applied (eg, 1.5 for 100 and 150 cm, respectively).

Detect a Ball Bearing Center

Some example tests presented are based on imaging the ball bearing phantom. The ball bearing produces a high contrast spot in a given test image and can be detected automatically or manually where robustness of the algorithm is insufficient. A script platform for manually entering the reference pixel coordinates is presented in the next session. Here, let us focus on a simple algorithm in order to detect a high contrast ball bearing image in a specific collimated small radiation field. An example of such configuration is shown in Figure 8.3.

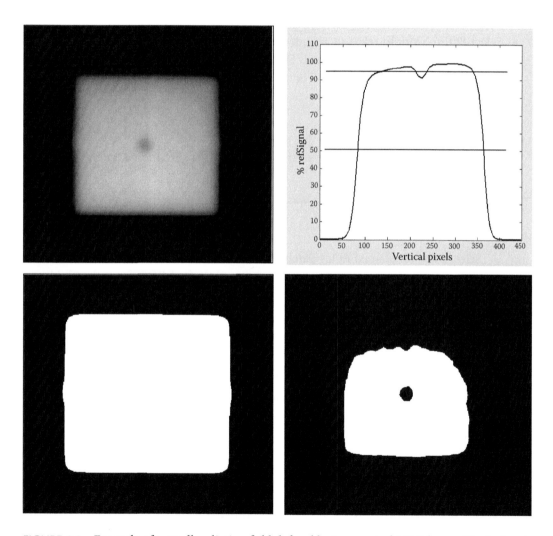

FIGURE 8.3 Example of a small radiation field defined by two central MLC leaves. The leaf ends form the left and right field edge, the leaf sides the top and bottom edge. The ball bearing image is inside the field (top left). Normalized signal profile from a contrast enhanced image (top right) shows discrimination levels to create the reference field mask (50%, bottom left) and the ball bearing mask (95%, bottom right).

Algorithm Scheme

Prerequisities

Test images: One test image in DICOM format imported in MATLAB, small square radiation field around the image of the ball bearing (Figure 8.3). The ball-bearing image is expected inside the reference field. The reference field should be large enough to prevent the ball bearing appearing close to the field border.

Algorithm Steps

- Select ROI to work in relevant area of the image only
  - Determine the reference signal. Considering the conditions this can be again a robust signal maximum based on the image histogram …

    ```
    [counts, binCentres] = hist (testImg (:), 100);
    refSignal = binCentres (100);
    ```

  - Find the image signal center of mass (CoM). This should be somewhere in the field area …

    ```
    [I, J] = find (testImg > 0.5 * refSignal);
    CoM = [mean(I), mean(J)];
    ```

  - Select a square-shaped ROI – safely beyond field borders in all directions to work with. This can be expressed in number of pixels (*pxlsROI*) from the *CoM* and depends on pixel resolution and field size at the detector plane.

    ```
    testImgROI = testImg (CoM (1) - pxlsROI:CoM (1) + pxlsROI,
                CoM (2) - pxlsROI:CoM (2) + pxlsROI);
    ```

- Interpolate 2D the *testImgROI* in order to increase the accuracy of the reference field edge detection (see the section '2D Data/Interpolation and Resizing' in Chapter 2)
- Create the field mask by selecting pixels with the signal above a given threshold derived from the reference signal

  ```
  refFieldMask = zeros (size (testImgROI));
  refFieldMask (testImgROI > 0.5 * refSignal) = 1;
  ```

  This mask represents the reference field borders for measuring asymmetry of the ball bearing position when this is determined.
- Contrast enhancement might be useful for MV images. Method and magnitude depends on actual conditions. For this example, a simple *contrastFactor* matrix is created with values depending on a given pixel signal difference from the mean

```
contrastFactor = (testImgROI/mean (testImgROI (:)) ^ 1.5;
testImgROIc = testImgROI.* contrastFactor;
```

- Create the ball bearing mask using another signal discrimination allowing detection of the ball bearing position as demonstrated in Figure 8.3. Calculate a new *refSignalC* first and then create the mask …

```
[counts, binCentres] = hist (testImgROIc (:), 100);
refSignalC = binCentres (100);
ballBearingMask = zeros (size (testImgROIc));
ballBearingMask (testImgROIc > 0.95 * refSignalC) = 1;
```

Setting appropriate signal discrimination levels is a crucial factor of this simple algorithm and depends on parameters of exposure. In the case of an MV beam, this is the quality of the imager and, of course, beam energy and number of MUs used for image acquisition. In the case of kV, this is again the detector and exposure parameters (kV and mAs). Instead of fixed nominal thresholds (0.5 from the original and 0.95 from the contrast enhanced image in this example), it is possible to consider alternative options such as the image histogram and setting the appropriate threshold for the ball bearing as center of the 98th bin of 100 calculated. It is also possible to consider calculating this histogram from the reduced field area by the first discrimination in order to increase the resolution. The final values for thresholds will depend on the test and equipment specific conditions and can be fine-tuned using specific local data.

- Determine the ball bearing position …
  The last step in determining the ball bearing center is to extract the ball bearing pixels from the *ballBearingMask*. To do this, it is efficient to use MATLAB's inbuilt function *imfill* which identifies and removes 'holes' in a binary mask image.

```
NoBBSpotMask = imfill (ballBearingMask, 'holes');
ballBearingMask = xor (ballBearingMask, NoBBSpotMask);
[BBSpotRows, BBSpotColumns] = find (ballBearingMask > 0);
```

The ball bearing position in matrix coordinates then calculates as the spot center of mass:

```
ballBearingPosition = [mean (BBSpotRows, mean (BBSpotColumns)]
```

Manual Input of Reference Pixel Indices

When a radiation test field is larger or more complex with more objects in it, automated detection might be more challenging. One possible approach is to search for an expected object (reference marker, ball bearing, field border) in the vicinity of the expected position in the image based on test parameters and previous experience. This can work well for tests where the data acquisition procedure is set rigorously and parameter variation is small.

TABLE 8.1   Example of a simple MATLAB script to input two pairs of reference points (ball bearings) per test image of total four stored in a cell array type variable *testImages*

```
for i = 1:4
    imagesc (testImages {i})
    colormap gray
    caxis ([CaxisLower CaxisUpper])
    disp ('Zoom on ball bearing 1. Press SPACE BAR when ready ... ')
    pause
    disp ('Use mouse to enter ball bearing 1 center position ')
    [x (2*i-1), y (2*i-1)] = ginput (1);
    disp ('Zoom out and in again on ball bearing 2. Press SPACE BAR when
 ready ... ')
    pause
    disp ('Use mouse to enter ball bearing 2 center position ')
    [x (2*i), y (2*i)] = ginput (1);
end
```

Alternatively, one can consider manual entry of reference points. A simple procedure can be split in few phases:

- Display the reference image including the cursor position and pixel signal value indicator (*impixelinfo*)

- Adjust the contrast window parameters using MATLAB's inbuilt function *caxis*

- Enter the reference pixel indices using MATLAB's inbuilt function *ginput*

An example of a simple MATLAB script to input two pairs of reference points (ball bearings) per test image of total four is illustrated in Table 8.1. The command sequence includes simple prompts displayed in the command window.

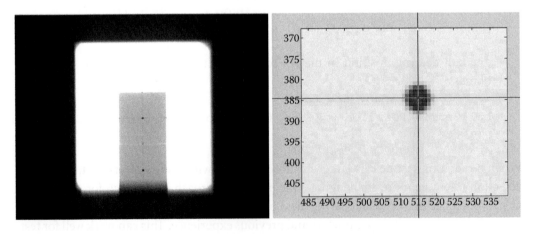

FIGURE 8.4   Example of a test image with manual input of the ball bearing position in matrix coordinates using the *ginput* function.

The output is an array x with matrix row indices for the ball bearing 1/test image 1, ball bearing 2/test image 1, ball bearing 1/test image 2, and so on. The y array is the column indices analogy to the array x. Example test image including the *ginput* interface is shown in Figure 8.4.

Knowing the simple principles to detect a beam collimation device edge and both automated and manual ball bearing image center from test images, we can now demonstrate examples of measuring a conventional radiotherapy accelerator mechanical parameters from specific test images acquired using the machine's standard imaging detectors.

MV Imager Essentials

To start with measurements of other parameters, the measuring device, the MV imager, vertical position, and orientation must be first verified. The imager plane is expected to be *perpendicular* to the radiation beam axis at the correct *vertical position* (source detector distance, SDD). Other degrees of freedom, for example, the detector position relative to the beam axis or rotation relative to the major anatomical directions, can be tested later when the beam axis, isocenter and reference directions are actually defined.

The detector perpendicularity and vertical position can be verified, for example, using measurements of the reference distance on a test image in two directions. Two fields defined by the MLC leaf sides with a nominal distance of 20 cm for nominal collimator rotations of 90° and 0° exposed on the imager must show the correct dimensions determined based on the nominal pixel size and the given SDD. The same discrepancy in both directions indicates a likely discrepancy in the imager distance from the source, longer dimension in one direction indicates a likely detector panel tilt in the given direction. The reference distances can be measured from test images reference profiles using the general approach described earlier, however, there is one important aspect to consider. As the imager horizontal rotation and position are unverified at the moment, it is not possible to extract simply a row and a column

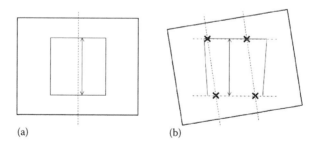

(a) (b)

FIGURE 8.5   Assuming both MLC leaf sides are parallel (horizontal field border in both a and b) and the detector orientation matching, it would be simple to measure the reference distance from a signal profile extracted simply as a column vector from the image matrix (dashed line in a). But because the detector can be rotated relative to the field collimation device, the reference distance (between the two MLC leaf sides) must be determined as the distance between two parallel lines (dashed-dot line in b), each defined by 2 points on the respective field border. The lines defining the points (in units of matrix indices) are determined from convenient column vectors extracted from the image matrix (dashed lines in b). Note the vertical field collimation is also considered neither parallel nor orthogonal to anything.

profile to measure the reference dimensions. A more general algorithm has to be followed in order to ensure a correct reference dimension is measured. See Figure 8.5 for details and the collimator spoke test algorithm description below with a similar approach applied.

Collimator Rotation Spoke Test

The key factor for using the sides of the MLC leaves for the definition of the radiation beam axis is the accuracy of the MLC lateral placement. The test verifying this parameter is the collimator rotation spoke test.

Assuming the imager vertical position and perpendicularity to the beam axis has been verified, we can proceed to the collimator rotation test.

Test Plan

Similar to the previous test, use the MLC to shape a test field. Just opening 2 or 4 central leaves (depending on leaf width) produces a strip-like field 1–2 cm wide. Leaves should be opened enough to provide some length for the test accuracy: the longer the field, the smaller the uncertainty in the determination of the parameters of the straight line defining the field axis. Depending on the imager dimensions, this can be 25 cm and possibly more. Secondary jaws should be opened so they do not intervene with the beam collimation. Create a minimum of four test fields with different collimator rotation angles sampling the whole range of motion, for example, 0°, 90°, 135°, 225°. The four test fields can be part of a single dedicated treatment plan in the TPS.

Data Acquisition

Acquire integrated images for each test field separately with the detector at the desired vertical position. Deliver sufficient MUs for a good quality image. Save or export the images in DICOM format.

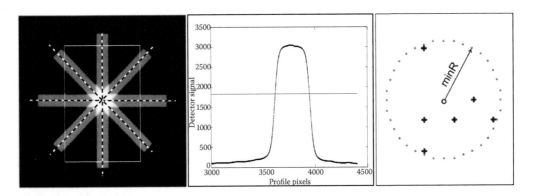

FIGURE 8.6 Example of a collimator spoke test for four different collimator angles producing 6 intersection points in total between four longitudinal field axes defined as midlines of MLC leaf-side collimated beams. All four test images are summed up but evaluated individually (left). In-field segment centers were found from the respective signal profile to provide two definition points for each field longitudinal axis. The minimum radius of the circle encompassing all intersection points, *minR*, is the parameter describing the collimator rotation walkout (right).

Data Analysis

Provided the test radiation fields are symmetric around the assumed radiation axis, the longitudinal field axis, that is, the straight line exactly in the middle of the two leaf edge images, crosses the radiation beam's central axis. The six (combination of four) intersection points then represent the radiation beam axis walkout due to the combined effect of the collimator mechanical rotation axis and beam collimation asymmetry. The example result is presented in Figure 8.6. The minimum radius of the circle encompassing all intersection points is the parameter describing the collimator rotation walkout. The radiation beam axis for any collimator rotation angle will be the maximum of this radius value off one virtual point which can be considered *the best compromise* collimator isocenter.

Algorithm Scheme

Prerequisities

Test images: Four test images in DICOM format showing strip-like radiation fields symmetrically collimated by central MLC leaf sides

Output:

    minR   The minimum radius of the circle encompassing all intersection points found

angleDeltas   Angular deviations from expected angles between each spoke and the reference spoke (90°)

Algorithm Steps

Importing DICOM images
- Import, rescale and convert test images in MATLAB 2D-array

Determining individual field longitudinal axes parameters
A straight line (longitudinal field axis here) is determined by two points:

- Loop for individual test images …
  - Interpolate the 2D test image by a factor of 4 (see the section '2D Data/Interpolation and Resizing' in Chapter 2)
  - Extract four test profiles along the sides of the rectangle shown in Figure 8.3. The rectangle is specific to the detector and field size and should be as big possible to increase the accuracy by measuring long rather than short distances but it must cross the fields before their longitudinal ends

Note: That for the given geometry, the rectangle is set in the way that for each individual strip-like field there are only two full signal (in-field) segments on opposite ends of the rectangle.

- Determine the reference image signal (*refSignal*) to find both in-field segment borders, eg, as the 100th bin center of the image histogram
- Loop for 4 test profiles …

– Check that the given profile contains the in-field segment by searching for signal above threshold, eg, using the condition …

```
currentProfile > 0.5 * refSignal
```

– For (two) positive profiles then find pixel indices of the centers of both in-field segments. These will be the averages of indices of two edge pixels of the given in-field segment. These two pairs of pixel coordinates define the given straight line (the given test field longitudinal axis)

At this point, there are four pairs of reference points, each defining a given test field longitudinal axis. From analytic geometry it is simple to derive the parametric equation in the matrix indices coordinate system for each straight line and calculate six intersection points between each two.

Calculating Field Axis Intersection Points   From analytic geometry, the intersection point of the two straight lines can be determined easily. See the associated MATLAB script (*colRot*) available in Web Supplement 8.1.

Determining Final Results and Output   The final output, the *minR*, requires searching for the circle center and the radius meeting the criteria. We are looking for a point with the minimum of the maximum distances between this point and all intersection points. This is clearly an optimization problem, but because of only six points that are close to each other, it is very simple to scan through the area defined by the minimum and the maximum row and column coordinates of the six points, calculate the maximum of a tested point to all six intersection points, and record the minimum value of all these maxima which is then the required *minR* parameter. To increase resolution and accuracy, the 2D interpolation of the search area is desirable. This non-elegant but functional approach to find the *minR* is used in the corresponding script (*colRot*) in Web Supplement 8.1.

Calculating Angle Deltas   A convenient side result of the analysis presented is the opportunity to calculate easy angular deviations between the spokes (test fields longitudinal axes). Defining one nominal angle as the reference, it is trivial to calculate the respective angle for all other spokes. Considering the MLC leaf side was defined as the primary reference for direction, the natural choice of the reference collimator rotation angle is 90°. It is trivial to verify this angle by another independent test (see the section 'Reference Collimator Rotation Angle 90°' in this chapter). The angle between the two straight lines with directional vectors *X1* and *X2* is calculated simply by:

```
rad2deg (atan (dot(X1, X2)/norm(X1)/norm(X2)))
```

The presented algorithm is independent of the imager rotation and position in the plane perpendicular to the beam axis. Only the panel vertical position should be correct to present the final result in millimeters (required for pixel to millimeter conversion).

Reference Collimator Rotation Angle 90°

The choice of 90° for the reference angle is natural for its simple validation and independence on other parameters such as gantry rotation scale.

Test Plan and Data Acquisition

A spirit level can be placed on the couch around the isocenter and levelled horizontally regardless of the couch. The top edge of the level represents the reference. Move the gantry to nominal 90° (does not need be accurate) and the MV imager close to the couch to acquire the image of the reference edge. The accuracy of the imager position is also not crucial. Open the secondary collimator and set the field shape using the MLC so that the leaf side is a few centimeters above the reference spirit level edge, and acquire the image where both the MLC leaf side and the reference level edge are a few centimeters apart.

Data Analysis

The test objective is to check the parallelism of both reference edges. This is easily done by comparing separations close to the longitudinal edges of the image. For such tests the units are not important so the distances can be measured naturally in pixels and the error angle determined from it. For best results, the field length together with the spirit level should be as long as possible to match longitudinal detector dimension. A definitely detectable difference of 1 pixel, for example, 0.392 mm, at the distance of 300 mm gives the corresponding angular error resolution of 0.07° *rad2deg(atan (0.392/300))* which is certainly acceptable from a clinical application perspective. See Figure 8.4 for a graphic demonstration.

Algorithm Steps

- Import, rescale and convert the test images in MATLAB 2D array using standard procedure described above

- Extract two signal profiles close to each longitudinal field edge and normalize them individually to their *robust maxima* to eliminate the impact of potential longitudinal field intensity or detector response variation

- Interpolate both profiles by the same factor to increase accuracy

- For each interpolated profile (*normProfile*) find pixel coordinates with normalized signal above threshold (80% in the example shown in Figure 8.7)

```
pxlsAbove80 = find (normProfile > 80);
```

- Find coordinates corresponding to pixels at correct profile edge, ie, corresponding to the leaf and spirit level edge, not the field edge. A vector of differences can be used efficiently to discriminate the first detected edge from the second in the example shown in Figure 8.7

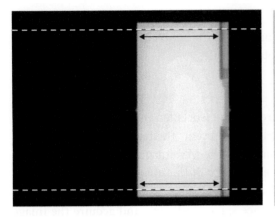
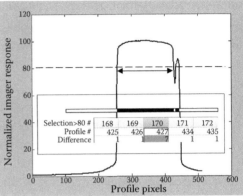

FIGURE 8.7 The test image with two signal profiles indicated. From the difference in distances between the MLC leaf edge at collimator angle 90° and visible edge of the reference spirit level, the corresponding angular deviation is calculated.

```
leafEdgePixelColumn = pxlsAbove80 (1);
spiritLevelPixelColumn = pxlsAbove80 (find(diff(pxlsAbove80) >1));
```

- Determine the distance between the detected edges (sufficient in pixels in this case) for both profiles and from their difference calculate the angle between the leaf and spirit level edges using the *atan* function

```
edgeToEdge = spiritLevelPixelColumn - leafEdgePixelColumn;

rad2deg (atan (difference-in-edgeToEdge [pxls]/distance-
    between-profiles [pxls]))
```

MV Isocenter and Reference Anatomical Planes

Provided the collimator rotation isocenter result described in the previous example is satisfactory, the collimator mechanical axis would be stable and the MLC used for beam collimation is sufficiently symmetric. The MV isocenter test presented in this example is, as in the case of the collimator rotation axis, determined on the basis of the radiation beam axis walkout in the whole range of gantry angle positions. Due to the beam axis definition adopted, the beam axis walkout due to gantry rotation will also include residual errors in the MLC placement relative to the collimator rotation axis and the collimator axis itself. Therefore, it needs to be decided whether or not to include the collimator rotation to the gantry isocenter test and, if so, how. In this example, the six test fields include collimator rotation for 0°, 90° and 270° for two reasons:

- It allows using only the MLC leaf sides for beam CAX definition for a given gantry angle

- It partially samples the clinical range of gantry-collimator angle combinations

Test Plan

The test plan consists of the six MLC square-shaped fields demonstrated in Figure 8.3. The MLC pattern is the same for all fields, the difference is only in the gantry-collimator angle combination: GNT0-COL0, GNT0-COL90, GNT90-COL90, GNT180-COL90, GNT180-COL0, GNT270-COL270. Two fields determine a given direction beam axis for testing superior, inferior, left and right directions, single fields for GNT90 and GNT270 provide test data for anterior-posterior direction.

Data Acquisition

The ball bearing cube phantom is placed on the couch top in the isocenter using the laser aligning system such that the reference markers indicate the projected ball bearing position on each of the phantom sides. The six test fields are then delivered and associated images recorded on the imager at reference distance, for example, 50 cm below the isocenter.

Data Analysis

The ball bearing position is detected on each test image using the algorithm described above. In the next step, the ball position relative to the reference field borders must be determined. Asymmetry indicates the beam axis distance in the respective direction from a fixed reference point in space – the ball bearing. For each two opposite directions it is possible to find the relative displacement of the ball bearing to reach the compromise position. This is *the best compromise isocenter*, the center of the isocentric ellipsoid defined by the radiation field axis walkout for the gantry-collimator combinations involved. Using the difference in beam axis positions relative to the ball bearing from the opposite directions, the isocentric ellipsoid respective dimension is also determined. See Table 8.2 and Figure 8.8 for graphical demonstration.

The isocentric ellipsoid dimensions represent the mechanical precision of the given linac isocenter for the parameters involved in the test. The ball bearing distance from the best compromise isocenter then shows the aligning lasers offset. Using just a few iterations, the ball bearing phantom can be placed with high accuracy in the best compromise isocenter and provide the reference for definitive adjustment of the lasers. However, a single reference point is insufficient for definitive setup of three orthogonal reference laser planes. Provided the planes are perfectly horizontal or vertical, the minimum two points per plane are still required. One point is the isocenter. The second point must be determined by other means. Considering the results of the tests done so far, it can be assumed that the

TABLE 8.2   Asymmetry in distance of given field borders from the ball bearing indicate the radiation beam axis deviation in a given direction from the reference point represented by the ball bearing

GNT-COL Combination	Asymmetry in Direction …	GNT-COL Combination	Asymmetry in Direction …
GNT0-COL0	Sup-inf	GNT180-COL90	Left-right
GNT0-COL90	Left-right	GNT180-COL0	Sup-inf
GNT90-COL90	Ant-post	GNT270-COL270	Ant-post

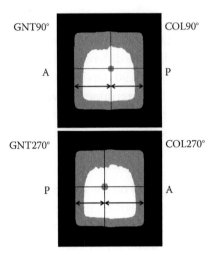

FIGURE 8.8 Determining the best compromise isocenter position in the anterior (A)-posterior (P) direction: the first image shows the field midline (beam axis), 0.2 mm anteriorly, and the second image, 0.4 mm anteriorly, from the reference ball bearing. The best compromise isocenter in the anterior-posterior direction is then 0.3 mm anteriorly from the ball bearing. Each image shows combined masks extracting the field border and the ball bearing, the calculation is based on separate masks (see Figure 8.3).

collimator rotation scales for 90° and 0° are verified so that the MLC leaf side can be used again, as the reference direction for major anatomical planes.

Let's place the second ball bearing cube on the couch at the maximum distance from the isocenter in the direction from the gantry that makes both ball bearings visible on the MV imager for the radiation field of appropriate size exposed from the gantry at 0° and collimator 90°. This is so that the MLC leaf side is parallel to the direction defined by the two ball bearings. Following the analogous process in the above examples (collimator rotation 90° test and manual ball bearing detection), the second ball bearing cube can be iteratively positioned until its distance from the reference MLC leaf side is the same as for the first ball bearing placed in the isocenter. At that moment, the reference markers on both phantom surfaces represent the two reference points for the reference sagittal plane. The ceiling laser can be adjusted to match the sagittal one in the whole covering range.

The reference transversal plane for both lateral aligning lasers can be defined analogously for the MLC leaf side and collimator rotation 0°. From now on, the aligning laser system can be considered correct and used as the secondary reference for the next tests.

kV Isocenter

With the MV isocenter test done knowing the best compromise isocenter position relative to the reference ball bearing, it is efficient to perform the kV isocenter test with the ball bearing cube untouched. Based on four orthogonal kV images of the ball bearing cube phantom it is easy to determine the kV analog deviation from the treatment, that is, MV, isocenter.

Test Plan and Data Acquisition

The test example consists in acquiring one image with the imager centered from each cardinal angle whether acquired manually or as a dedicated clinical plan.

Data Analysis

The clinical application of the kV system centered assumes a point object placed at the isocenter to be displayed on the kV imager central pixel (digital crosshair) from all directions and, of course, correct dimensions. Discrepancies detected are due to a combination of incorrect detector and X-ray source positions. Due to the closer distance of the detector to the isocenter, the image is more sensitive to errors in position of the detector.

The four test images can be analyzed using the basic procedures described above – manual ball bearing detection. The procedure is analogous to determining the MV isocenter with the difference that instead of beam axis defined based on a given symmetric field midline, in the case of kV, it is the central pixel. So, for each cardinal direction the central pixel deviation from the ball bearing can be determined. Knowing the relative position of the best compromise MV isocenter from the same ball bearing it is easy to calculate the kV axis deviation from this clinically most relevant point. In general, there are few options how to quantify or report the kV test result:

- Distance in mm of the kV axis from the MV best compromise isocenter projection for each cardinal direction

- (3D) distance in mm of the kV best compromise isocenter from the MV best compromise isocenter

- Distance in mm of the kV axis from the MV beam axis from the same direction, for all cardinals again

The first option is the most clinically relevant, and so is preferred. Considering the offsets measured are stable and characteristic for a given direction, it is possible to account for this in a given treatment machine's image analysis software and eliminate the IGRT matching error component due to this systematic factor. This includes correcting the projection images for CBCT before reconstruction.

MV Detector, kV Detector (kVD) and kV Source (kVS) Arms

Based on the tests presented so far, we have the reference point (best compromise MV isocenter) represented by one ball bearing cube phantom, and the reference direction, given by the second ball bearing cube phantom placed in the sagittal or transversal plane (see the section 'Image-Based Approach/MV Isocenter and Reference Anatomical Planes' earlier in this chapter). In addition, one can expect that the distance between the two ball bearings is known with high accuracy.

Following the same principles of the previous tests, it is then easy to verify the kVD-kVS geometry. The ball bearing indicated directions must correspond with the detector rows or columns for all vertical positions. The horizontal (perpendicular to the X-ray beam axis) alignment of the detector is based on the central pixel vs. the isocenter ball bearing. The vertical scale is verified using the reference distance measurement and nominal pixel size in the whole vertical range.

The remaining degrees of freedom of the MV detector can be verified using exactly the same principles.

Gantry Rotation Scale
Test Plan
The same test fields for the MV isocenter test can be used since this example test consists of acquiring MV images of the ball bearing cube phantom. There are two exposures for each of the cardinal angles. The exposures differ in the ball bearing cube phantom distance from the source in the given beam direction.

Data Acquisition
Obviously, the treatment couch translations can be used to transfer from one reference position of the phantom to another. However, as the couch motion may not yet be 100% accurate along the given direction, the alignment lasers play their secondary reference role here. So, for the gantry angles 90° and 270°, the couch with the phantom needs to be moved laterally, but the phantom must be corrected as per the reference laser when this indicates imperfect motion in the given direction. The alignment laser system has been set and verified before, together with the MV isocenter and the defining major anatomical planes tests. It is essential that the gantry + imager do not move between acquisitions of the two related test images.

Data Analysis
Provided the given test image pair was acquired correctly, it is trivial to determine the gantry rotation angle deviation in both radial and transverse (digital scale) directions, using the automated ball bearing detection described above. This is followed by calculating the corresponding angular discrepancy based on the difference in the ball bearing positions in the respective image, and the known phantom translation for the given gantry angle.

Couch Translation Orthogonality
The (conventional) treatment couch translates in vertical, longitudinal and lateral directions. These must agree with major anatomical directions determined using the reference set (MLC leaf side, MV isocenter, collimator reference angle 90°, collimator rotation scale) and indicated by the alignment lasers. This test probably does not require quantification as visual verification that a given test object placed on the couch moves in agreement with the given laser along the whole range of given motion and is arguably more sensitive than the image-based approach applied otherwise. However, from the observed deviations (Δx) and known distance (d) of given translation it is trivial to determine the corresponding angular deviation again when required: *rad2deg (atan ($\Delta x/d$))*.

Couch Translation Scale

The same general principles apply. The ball bearing cube phantom placed on the couch, image 1 acquired, the couch relative motion, image 2 acquired, etc. The difference between nominal (digital scale) and real distance is determined from the known detector nominal pixel size and the detector vertical position.

Couch Rotation Scale

The test analogy to the collimator rotation scale spoke test is described previously. The only difference is that the 'spokes' are no longer narrow radiation fields, but two ball bearing phantoms placed (and fixed) on the couch. The line connecting the two ball bearings should cross the best compromise MV isocenter in order to eliminate this component in the test result (analogous to the radiation field asymmetry in the collimator test equivalent). Then again four couch rotation angle sampling the range of motion, for example, 0°, 45°, 90°, 135°, four test images acquired, manual or automated detection of both ball bearings, parametric equations of the respective straight line, six intersection points and finally, the minimum radius of the circle encompassing all intersection points. A convenient side result: true angles between each spoke and the reference (eg, 0°) compared with nominal angles applied during the test.

Secondary Jaws (X, Y) and MLC Tests

Secondary jaws position scales can be tested again for the series of positions representing the whole range, for example, 125 mm, 50 mm, –10 mm (X)/–100 mm (Y). Negative positions, possibly different for the X and Y jaws due to their dynamic wedge capability, should also be included. Referring to the primary reference definition for the whole concept, the reference point, the origin, or beam central axis point, should be defined using the MLC. Acquiring an image with a convenient MLC pattern enables *pixel-to-mm from the CAX* calibration which can be then used for the jaws position measurements. The calibration can be obtained from the image signal profiles through the area of multiple leaf sides providing the calibration data [row/column, mm]. Using this calibration and general principles described previously it is then easy to detect the respective jaw edge and convert the given pixel coordinate in millimeters from the CAX.

In principle, there are two options to get the MLC calibration data in the direction of the leaf travel. Relying on the leaf positional calibration, it is possible to obtain the calibration data from the position of leaf ends. Alternatively, keeping the primary reference principles of the concept, one can rotate the collimator by 90° and derive the calibration data from the leaf side positions again. It depends on whether we prefer uncertainty in MLC leaf end calibration, or the collimator rotation axis included to the calibration data. The example scripts available in Web Supplement 8.1 provide results considering both approaches. Of course, in theory it is entirely possible to use the projection of the best compromise MV isocenter to represent the beam radiation axis reference on the imager and apply the corresponding correction to the MLC *pixel-to-mm from the CAX* calibration data.

By applying the collimator 0+90/MLC leaf side calibration, it is also simple to measure the position of the MLC leaf ends and verify the MLC calibration itself. Dedicated MLC patterns can be applied to test this efficiently.

The MLC performance in dynamic mode and the use of integrated images is based on rather final product in terms of uniformity of given areas exposed, and so on. Certainly, using the general principles presented in this chapter, one can create efficient scripts to analyze these tests too.

Example test images acquired on a clinical treatment machine are available in Web Supplement 8.2.

REFERENCES

AAPM TG100. (2016). The report of Task Group 100 of the AAPM: Application of risk analysis methods to radiation therapy quality management. *Medical Physics* 43(7), 0094–2405.

AAPM TG142. (2009). Task Group 142 report: Quality assurance of medical accelerators. *Medical Physics* 36(9), 0094–2405.

IPEM. (2006). *Acceptance testing and commissioning of linear accelerators*. United Kingdom: IPEM.

CHAPTER 9

Comparing Dose Distributions

The Gamma Method

LEARNING OUTCOMES

LO 9.1   Define the terms used when comparing dose distributions
LO 9.2   Discuss the importance and applications of comparing dose distributions
LO 9.3   Explain the basic concepts of qualitative and quantitative comparison approaches
LO 9.4   Explain the principles of comparing dose distributions using gamma analysis
LO 9.5   Explain the proposed MATLAB scripts for calculating the gamma map for two given dose distributions
LO 9.6   Design own script based on individual preference
LO 9.7   Use the MATLAB script to calculate the gamma map for two given dose distributions and interpret results
LO 9.8   Discuss the merits and limitations of the gamma analysis and consider alternative approaches to the comparison of dose distributions

INTRODUCTION

This chapter provides a guide to implementing the widely used method for quantitative comparison of two dose distributions in radiotherapy physics: the gamma method. At first, the general aspects and rationale for comparing dose distributions in radiotherapy are introduced. This is followed by the examples of natural, first choice, parameters including explanations of their disadvatages leading to creating the concept of the gamma evaluation. The main part of the chapter consists in describing the basic gamma algorithm and its variations including impact of these alterations on the final result. The last part of the chapter deals with discussion on various ways of results presentation and clinical relevance and robustness of the gamma method itself.

IMPORTANCE OF COMPARING DOSE DISTRIBUTIONS IN CLINICAL RADIOTHERAPY

There are often instances when one needs to compare two dose distributions in clinical radiotherapy. Examples are TPS commissioning, pre-treatment plan verification, treatment machine and software upgrades. This chapter will illustrate the methods used by considering the case of TPC commissioning. In this case one compares the dose distribution given by the TPS (here referred to as the 'reference distribution') to that of the actual measured distribution (here referred to as the 'evaluated distribution').

One of the major responsibilities of a radiotherapy medical physicist is TPS commissioning. The main function of a TPS is calculating the spatial dose distribution expected within a patient model resulting from simulated irradiation of the patient by a composition of radiation beams with various parameters such as nominal energy, direction, size, shape, weight and fluence modulation. The set of radiation beams and associated parameters resulting in best compromise clinical objectives are found using a combination of TPS dose calculations and appropriate optimization algorithms. Some modern TPS include radiobiological metrics in addition to physical absorbed doses.

In brief, the commissioning process starts with the inputting into the TPS of essential measured, processed and/or derived beam parameters in order to provide source data for the beam model. Beam data is determined by using dedicated radiation measurement systems in water. All modern TPS are (at least) three dimensional, with dose calculations performed in 3D on patient voxel models. The patient voxel model is derived from a planning CT although MRI based patient voxel models are on the increase. Once a particular TPS is configured and capable of producing calculated dose distributions, radiotherapy physicists must verify that the calculated dose distribution is in agreement with actual absorbed dose measurements.

Basic verification is carried out by comparing the dose distributions produced by single beams of known parameters on a water equivalent model and calculated using the TPS with corresponding measured values. For the most commonly used high energy photon beams, single beam dose distributions such as dose profile along central beam axis for various field sizes, tissue phantom ratio or lateral beam profiles that are perpendicular to the central beam axis produced by the TPS can be verified in this rather straightforward way. However this simple verification is trivial since it is linked directly to beam data inputted into the TPS as part of the commissioning process.

A more complicated situation arises with beams with parameters not measured directly for TPS input during the commissioning process. Examples of such beams are wedged beams, MLC-shaped beams and, of course, fluence-modulated beams. Wedged beams can be verified by comparing relevant beam profiles perpendicular to the central beam axis at given depths. However, it is clear that at some point multi-beam and multi-dimensional dose distributions must be considered.

Basic beam verification for a single beam in 2D is typically done using a planar detector/dosimeter with the detector plane normal to the beam axis and with some material above and/or below the detector, depending on the objective. Alternatively an EPID can be used to record the measured dose/response perpendicular to the beam axis to compare it with the corresponding calculation by TPS.

Nevertheless, the verification process should also include *end-to-end* tests with clinical multiple-beam treatment plans resulting in complex dose distributions. These clinical-like dose distributions are verified by a suitable 2D dosimetry system in combination with an appropriate phantom chosen according to the test objectives.

There are many radiation detectors for performing 2D dose measurements available today. The major three categories are:

- Electronic detector arrays (single-planar or multi-planar or cylindrical)

- EPID, as a special case of category one (single-planar, perpendicular to beam axis only)

- Films (today mostly radiochromic, self-developing)

In a similar way, 3D doses can be verified. In such a case, it does not make much sense to verify single beams, so typically, integral 3D doses of composite treatment plans are the subjects of comparisons. Three-dimensional doses are calculated by a TPS on dedicated phantom models. These phantoms are then irradiated on a treatment machine and the respective 3D dose spatial distribution is measured directly or indirectly. Direct 3D dose measurements with spatial resolution comparable with the resolution of dose calculation can be done by chemical *gel* dosimeters or their equivalents. Alternatively one can think of a stack of 2D dosimeters measuring dose in multiple layers of a given phantom. Indirect measurements are based on 3D dose reconstruction from 2D projection data. 2D projections can be acquired either with or without a phantom/patient in the beam paths. The latter case is known as *transit dosimetry*.

Regardless of detector type, it is always the measured dose distribution that is compared with the corresponding TPS calculation. From a mathematical perspective, an important issue is the difference between the spatial resolutions of the reference and the evaluated dose distributions.

There is a substantial difference between the problem of comparing a single uniform or even a flattening-filter-free (*FFF*) open beam and a fluence-modulated beam or multiple (fluence-modulated) beam dose distributions. The nature of the latter two is complex with the presence of high dose gradients to meet the general goals of modern radiotherapy that is concentrating high dose on target while minimizing dose to critical organs. Modern radiotherapy technology such as fully exploited IMRT/VMAT has the potential of highly complex dose distributions created with precisions of the order of a millimeter.

QUALITATIVE AND QUANTITATIVE APPROACHES

In general, dose distributions can be compared both qualitatively and quantitatively. Qualitative comparisons are subjective, typically visual, without quantitative metrics and simply in terms of higher or lower absorbed dose values. In terms of dose distributions, the simplest approach is isodoses or dose profile overlays. With qualitative comparison there is always a question of whether or not the observed level of agreement is satisfactory or otherwise. See Chapter 2 for an introduction to basic dose 2D-array operations and isodose display.

The major feature of quantitative assessments is their objectivity and direct evaluation by comparison with preset recommended reference tolerance levels. There is a whole history of methods of quantitative comparison of two 2D dose distributions including rec-ommended tolerances (eg, ESTRO 2008). The simplest parameter for assessing the level of agreement of two 2D dose distributions is relative dose difference ΔD:

- Let's assume two same size 2D-arrays A (evaluated) and B (reference) represent-ing two 2D dose distributions with identical spatial resolution to compare using a simple 2D-array subtraction – dose difference 2D-array. The relative dose difference is given by

$$\Delta D_{ij} = (A_{ij} - B_{ij}) / B_{ij} \times 100\% \qquad (9.1)$$

- Having a relative dose difference 2D-array one can easily obtain the fraction of all elements in the subtraction 2D-array with values lower than a preset tolerance of, say, 5%

- Such fraction (of points with relative dose difference below 5%) is an objective metric and can be used as a single quantitative parameter reflecting the level of agreement of two dose distributions

A problem with this metric is that it gives high difference values in high dose gradient areas even in cases with only a small geometric misalignment. A perfect geometric align-ment of the measured and the reference dose distributions is difficult to achieve in prac-tice, not only because of the finite mechanical accuracy of technology and setup, but also because of the possible different spatial resolutions of the TPS and the dosimeter. To avoid this problem, the *distance-to-agreement* (DTA) concept was introduced (Venselaar 2001). DTA is a geometric distance d from a point in the measured distribution to the closest reference distribution point with equal dose.

Tolerance criteria for quantitative assessment can then be set for the various regions of interest (ROIs) of the dose distribution for example:

- 3% dose difference for high dose and low dose gradient areas

- 5% dose difference for low dose and low dose gradient areas

- 3 mm DTA for high dose gradient areas

Provided the high/low dose gradient areas are clearly distinguished (eg, >/<5% per 2 mm), it is mathematically simple to assess all tested measured dose points against the reference dose distribution and produce basic statistics, and so on.

Theoretical Foundations of the Gamma Method

The current widely acknowledged standard method for comparing two dose distributions is the 'gamma analysis' introduced by Low et al. (1998). Basically, instead of using dose difference and DTA criteria separately as described above, it combines both metrics into a single 'gamma index' given by:

$$\gamma(r_r) = \min_{\forall r_e \in ROI} \{\Gamma(r_r, r_e)\},$$

where (9.2)

$$\Gamma(r_r, r_e) = \sqrt{\frac{r^2(r_r, r_e)}{\Delta d^2} + \frac{\delta^2(r_r, r_e)}{\Delta D^2}}$$

This chapter will present a basic implementation of the gamma analysis and demonstrate key assumptions and decisions one has to take in order to achieve this task. It is the number and complexity of these assumptions and decisions that makes the gamma method difficult to implement accurately per its definition.

Keeping to the same notation as that in Low et al. (1998), the Γ for each point r_r within the ROI of the reference dose distribution is calculated where r_e refers to a general point in the evaluated dose distribution, and Δd and ΔD are preset tolerances on DTA for distance $r(r_r, r_e) = |r_r - r_e|$ and dose difference $\delta(r_r, r_e) = D(r_r) - D(r_e)$ between the points r_r and r_e, respectively. The resulting γ value is interpreted as:

- if $\gamma < 1$ then $\delta < \Delta D$ AND $r < \Delta d$

- if $\gamma > 1$ then $\delta > \Delta D$ OR $r > \Delta d$

The current reporting standard of the result is the fraction of the points with $\gamma < 1$, that is, $P_{\gamma<1}$ for a respective pair of Δd and ΔD preset tolerances.

A standard application of the method is that values of the two tolerance parameters (dose difference tolerance, ΔD, and distance-to-agreement tolerance, Δd) are preset; software then calculates a corresponding 2D-array of γ values. Points with $\gamma > 1$ are then interpreted as failing preset tolerances and vice versa. The analysis returns the fraction of the points within a certain ROI passing the test ($P_{\gamma<1}$). A more complex result representation is the γ-histogram (Stock et al. 2005). The γ-histogram representation allows determination of histogram-derived parameters, such as maximum or mean γ and fraction of ROI with γ over a specific value, for example, $P_{\gamma>1.5}$, $P_{\gamma>2}$, and so on (De Martin et al. 2007).

For application of the method in clinical practice, in addition to the graphical result presentation, acceptance or action levels for test passing fraction of the ROI and/or other parameters of relevance are set. Based mainly on statistics of a large number of verified fields, recommended minimal fractions of ROI passing the test for respective values of basic tolerances, ΔD and Δd, have become subjects of guidelines for IMRT clinical practice (ESTRO 2008; IPEM 2008).

Implementation Assumptions, Conditions and Decisions

When implementing the gamma method there are several decisions to make. These decisions are both mathematical and clinical by nature and impact on the result. None of them is addressed directly by the basic principles and definitions described previously. As always, further decisions are required in order to optimize calculation speed and, again, some of these decisions may under certain circumstances impact the overall result. With respect to the gamma implementation, the following areas are key factors that need to be considered.

Data Import Formats

The algorithm inputs for a 2D problem are two generally rectangular 2D-arrays forming the dose distribution to be evaluated and the reference dose distribution, respectively. Typically, the dose distribution to be evaluated is exported from a measuring device in device specific format and the reference dose distribution is exported from TPS, usually in DICOM RT format. See Chapter 1 for examples of data imported into MATLAB.

Dose Normalization

Whether done before dose export from the measuring device or TPS, or later, dose normalization is an important factor in the entire analysis. Assuming both 2D-arrays are imported as absolute doses in Gy, the key question is normalization of dose difference δ because this quantity will always be compared with preset tolerance ΔD typically given in *nominal* percentage (eg, 3%). Dose difference normalization can be done in many ways with significant impact on the result:

- *Global* normalization of the dose difference

  Dose difference δ for all pairs of experimental/tested and reference points from respective 2D-array are normalized using a *single* constant value, for example,

  $$\delta(r_r, r_e) = \left[D(r_r) - D(r_e) \right] / D_r^{\max} \times 100\% \tag{9.3}$$

  where the maximum dose from the reference dose distribution is the global normalization constant. This role can be played by any other dose value from either dose distribution which is independent of both r_r and r_e.

  As a special case of the global normalization approach one can consider dose distributions normalized by separated constants but still independent of both r_r and r_e, for example,

  $$\delta(r_r, r_e) = \frac{D(r_r)}{D_r^{\max}} - \frac{D(r_e)}{D_e^{\max}} \times 100\% \tag{9.4}$$

  where the maximum dose from each respective dose distribution is a global normalization constant for that distribution giving dose difference again as a percentage of the normalized dose values.

  Choice of normalization constant for global normalization of dose difference in gamma analysis is arbitrary and specific to the situation and also to the user's preference. This is one of the major parameters that make gamma analysis complicated to standardize, generalize and share among various applications, situations, devices and institutions.

Reference dose maximum D_r^{max} is probably the most common choice for global normalization in gamma analysis. Preference of the reference over the experimental/tested dose is just to eliminate the risk of the potential undesirable impact of a single point dose accidentally measured wrongly. Justification or motivation for the choice of maximum dose is rather primitive: some decision needs to be taken and the point dose maximum is a clearly defined parameter. Another 'advantage' is that of all other possible options, this one is probably the 'softest' and leads to the smallest dose difference to be compared with the preset tolerance.

- *Local* normalization of the dose difference

The dose difference δ for all pairs of experimental/tested and reference points from respective 2D-array are in this case normalized using a variable, position dependent, normalization value, that is,

$$\delta(r_r, r_e) = \frac{D(r_r) - D(r_e)}{D_r'} \times 100\%$$

or (9.5)

$$\delta(r_r, r_e) = \frac{D(r_r) - D(r_e)}{D_e} \times 100\%$$

This means that all dose differences are normalized locally either by the respective reference or the experimental/tested dose. As such, the local dose is generally smaller than the respective dose maximum, so that δ values are generally higher than when global normalization is applied. This makes locally normalized gamma methods harder to pass. Whether to use the reference or the evaluated dose for local normalization depends, again, on user preference. As demonstrated later in this chapter, for resolution aspects (reference dose resolution should be small compared to DTA tolerance), choice of reference local dose $D(r_r)$ for the purpose of dose difference normalization may be preferred. On the other hand, if the impact of resolution is insignificant then the evaluated local dose $D(r_e)$ may be preferred due to the fact that it remains constant during individual evaluated point gamma calculation, that is, only dose difference and mutual distance to a reference dose point determine the value of the Γ for a given evaluated point.

In principle, with the same algorithm it is possible to compare both non-normalized dose 2D-arrays and relative doses pre-normalized globally. One should be aware that for normalized dose 2D-arrays the locally normalized gamma calculation result will be the same as for non-normalized ones only if the same normalization factor was applied to both 2D-arrays.

Global normalizations can be and should be done beforehand. The reason is that doing so spares one numerical operation in the calculation of the Γ values. Local normalizations must be done individually for the Γ calculation of each two relevant 2D-array elements.

Spatial Resolution

Spatial resolution also has an impact on the quantitative results of the gamma method. There are two aspects related to spatial resolution one has to consider for algorithm development. Spatial resolution of detectors varies and TPS dose distributions can be calculated and/or exported with various spacing.

As in the case of global data normalization, spatial resolution generally may be handled before import in MATLAB as both TPSs and the control software of measuring devices typically enable this.

The choice of resolution of the measured dose distribution is entirely arbitrary and up to the user's preference. The author's preference is to keep the detector nominal resolution. If the evaluated dose comes from a detector with arbitrary resolution, such as gel or film, the author's preference is to choose a clinically relevant value, typically 1 mm, isotropic. If the imported evaluated dose requires a change of resolution, then the algorithm will be the same as for the reference dose that is discussed later on.

Unlike the evaluated dose, the spatial resolution of the reference dose is important for the gamma method. If it's 'too coarse' relative to DTA tolerance (typically 3 mm) and local dose variation (gradient), then if the algorithm implementation does not take this into account there is a chance of an increased number of failing points affecting the overall result. There are at least two ways to handle this problem. The easiest ways are:

- Adjusting the reference dose spatial resolution adequate to given DTA tolerance and local reference dose variation, or …

- Incorporating interpolation in the gamma calculating algorithm

Naturally, both approaches are associated with increased computing demands.

Threshold

As mentioned earlier, for the gamma method the pass/fail fraction of evaluated points within certain ROIs is typically used as the final outcome (note that whether area or volume, how to define ROI is another arbitrary decision by implementer or a user). Instead of using a fixed or arbitrary ROI it is more clinically useful to compare high dose ROIs with therapeutic intent and/or potential of radiation injury of healthy tissue. Using a dose threshold with only points with doses above a certain level being evaluated, to define such ROIs provides some advantages with respect to other options (eg, whole detector area/volume) as any dose-to-dose comparison in terms of percentage difference leads to problems when the compared doses are small simply because of division of small numbers. In addition, uncertainty in both measurements and calculations increase at low doses. Unlike 'fixed' ROI, the ROI defined using dose threshold is also dose distribution size/volume independent. These are all reasons why it is common to exclude doses below given threshold from quantitative dose comparison.

In terms of implementation, threshold dose defined ROI is not as trivial as it might sound. Surely, for a globally normalized evaluated dose it is easy to select for evaluation only points above x-% of normalization dose. For the local normalization approach this is more complicated since no global normalization is required for the gamma calculation. Therefore, the dose threshold value to define ROI is an extra parameter, rather independent of the gamma calculation itself. If set in percentage dose then normalization (for this purpose) needs to be considered too. Simple and natural options are, again, evaluated D_r^{max}, reference D_e^{max} or a prescribed dose. (Relative) dose thresholds used commonly in practice are 10% or 20%. Whatever approach is preferred, ROI (threshold) choice has a direct impact on the gamma statistics.

Pass/Fail Only or Exact Gamma?

An important decision is whether it is necessary to know the statistics of exact gamma values (eg, mean γ) or whether a simple pass/fail fraction is sufficient for a given application. Accurate gamma values are required when any gamma statistics beyond pass/fail is considered for the test definitive outcome.

However, as gamma inherently includes intrinsic tolerances in clinically meaningful quantities, that is, dose and DTA, incorporating further complexity and moving from simple pass/fail towards extensive use of statistics may be considered unnecessary or counterproductive. Additionally, the simple pass/fail approach does not require searching the minimum Γ as per definition and accepts any Γ < 1. This makes the algorithm quicker too.

On the other hand, there are applications utilizing exact gamma distributions. One of them is that any gamma algorithm can be tested based on the assumption that minimum Γ has been subjected to search as per definition. For details, see the section 'Merits and Limitations of the Gamma Analysis'.

EXPLANATION OF THE PROPOSED MATLAB SCRIPTS TO CALCULATE THE GAMMA INDEX FOR TWO GIVEN DOSE DISTRIBUTIONS

The basic algorithm, in the form of a structured text, and linked to the respective MATLAB script in Web Supplement 9.1 is described in the next section.

Algorithm Scheme

- Function Syntax

```
myGammaSimple (doseEvaluated, doseReference, deltaDose,
deltaDistance, Threshold)
```

- Function Input/Output
  - Input
    doseEvaluated: Evaluated 2D dose distribution, rectangular 2D-array, resolution 1 × 1 mm², normalized globally, aligned with *doseReference*
    doseReference: Reference 2D dose distribution, same size 2D-array as *doseEvaluated*, resolution 1 × 1 mm², normalized globally, aligned with *doseEvaluated*
    deltaDose: Dose difference tolerance ΔD in %

deltaDistance: Distance-to-agreement tolerance Δd in mm

Threshold: Threshold level for determining ROI for gamma calculation in %
and *doseEvaluated*

- Output
  - 2D-array of the gamma values of the same size as the *doseEvaluated* with negative ones at non-evaluated points to exclude them easily from the statistics
  - Figure displaying the gamma 2D-array with dedicated gamma colormap: white for non-evaluated points, grayscale (white to black) for gamma from 0 to 1, and red for $\gamma > 1$
  - On-screen display absolute number of test failing points
  - On-screen display pass fraction of all evaluated points
- Preparations
  - Set calculation time meter
  - Pre-define the gamma 2D-array of the same size as the evaluated dose distribution with negative ones
    - Gamma values are calculated for each eligible point and original value is replaced by the corresponding gamma. This clearly distinguishes non-evaluated points as all calculated gamma are non-negative by definition
- Loop: Calculating gamma for selected evaluated points
  - For each row (i) and column (j) indices of the evaluated dose 2D-array ...
  - Check the evaluated dose *doseEvaluated (i, j)* is larger than *Threshold* dose. Do nothing if not. If so then ...
  - Calculate the first Γ_1 value for the reference point at identical position as the evaluated point, ie, the same (i, j) of the two registered 2D-arrays. This first Γ_1 determines 'mathematically meaningful searchspace' for the minimum gamma (γ) of the entire reference 2D-array *doseReference* as the *k_radius* factor equals to *floor* ($\Gamma_1 \times deltaDistance + 1$) ... see Figure 9.1
    - Mathematically, it is pointless to search for smaller gamma with reference points located beyond the value of the first (original) gamma multiplied by the *deltaDistance* tolerance as distance only contributions to the gamma vector for such points are larger than the first gamma Γ_1 so they can only be larger
    - Secondly, regardless of the first gamma the definition of the *k_radius* makes its value >=1 so immediate vicinity (ring), ie 8 neighbouring points, will always be included in search of a smaller gamma in order to avoid algorithm stopping with the first gamma <0.5 when there may be points around with gamma even smaller. The *k_radius* is constant during the minimum Γ search for each evaluated point and serves as an efficient speed up factor to eliminate a pointless search through the entire reference dose 2D-array
  - Loop: Searching through the meaningful search space defined by *k_radius*
    - See Figure 9.1 ... in the first loop search for smaller Γ among reference dose points with either row or column indices differing by 1 from the evaluated point (i, j). These points form a 'rectangular perimeter' around the (i, j) position

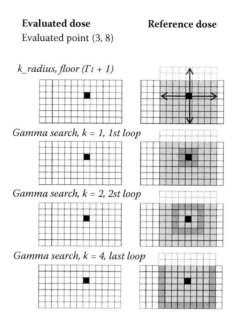

Evaluated point (3, 8)

Reference dose

k_radius, floor (Γ1 + 1)

Gamma search, k = 1, 1st loop

Gamma search, k = 2, 2st loop

Gamma search, k = 4, last loop

FIGURE 9.1 Demonstrating the definition of 'mathematically meaningful search space' for the minimum Γ and gradually increasing the search 'ring' perimeter including out-of-border test for one evaluated dose point.

- – … See the code for how to address only these points avoiding points with both index increments in absolute value smaller than desired 'perimeter' search space
- – Each addressed point must be checked for whether or not it lies within the reference dose 2D-array, ie, whether or not the new global indices are still within the respective 2D-array dimensions. Without this condition the algorithm would crash for the evaluated point positions closer to the 2D-array edge than respective meaningful search space radius
- – If the Γ is smaller than current smallest then replace the current smallest
- – See Figure 9.1 … in the second loop search for smaller Γ among reference dose points with either row or column indices differing by 2 from the evaluated point *(i, j)*
  - – The same conditions and notes as for the first loop apply
- – See Figure 9.1 … carry on through the entire meaningful search space (defined by *k_radius*)
  - – The same conditions and notes as for the first loop apply
- – Carry on the loop for all eligible evaluated points
- • Statistics
  - • Within the result gamma 2D-array calculate the number of non-evaluated points (original γ = −1)
  - • Within the result gamma 2D-array calculate number of test failing points (γ > 1)

- Results display
  - Absolute number of failing points
  - Pass fraction of all evaluated points
  - Gamma map: Non-evaluated points in white, points with γ between 0 and 1 incl. in grayscale, points with $\gamma > 1$ in red
    - This requires a dedicated colormap definition ... see the *gammaColormap* script in Web Supplement 9.1 and the colormap itself in Web Supplement 9.2

      The simplest way to define a colormap with desired properties is to take the original MATLAB *gray* colormap and modify it. The *gray* colormap is a 101×3 array with the first row of $[0, 0, 0]$ and the last row of $[1, 1, 1]$ indicating pure black and pure white color, respectively. Flipping the array upside down and replacing the $[0, 0, 0]$ (now the last) row with $[1, 0, 0]$ we ensure the last bin *matrix-to-display* values shows in red while the others in 100 levels of gray scale. This is not exactly what we want as for the gamma 2D-array display we require a sharp transition at the gamma value of 1. We can help this arbitrarily simply by increasing levels in the colormap from 101 to, for example, 256. This is the case of the *gammaColormap* variable in Web Supplement 9.2 which is satisfactory for normal use. In addition, in Web Supplement 9.1 there are two simple scripts available that display the calculated gamma map accurately. The *displayGamma101* and *displayGamma201* show the gamma map with a limited number of gray levels making the image appearance darker or lighter, with red indicating points with $\gamma > 1$ strictly, leaving no inaccuracy in display to a colormap bin resolution.
  - Total calculation time (information purpose only)

Alternatives to the Basic Gamma Calculation Algorithm

The following alterations address various aspects of the basic algorithm described earlier with the aim to further deepen understanding of the available options and decisions.

Calculating Gamma Using Norm

Usually, it is quicker to use MATLAB inbuilt functions as often as possible. Looking at the general definition of the gamma (see Equation 9.2) it is actually a vector norm in two-dimensional Dose $(D) \times$ Distance (r) space where the Distance (r) consists of further 2 sub-dimensions, X and Y (for 2D dose distributions), making the gamma space actually three-dimensional Dose $(D) \times$ dimension $X \times$ dimension Y.

The only differences in respective MATLAB code are in the gamma calculation formula:

```
sqrt ((doseEvaluated (i, j) - doseReference (i+kki, j+kkj))^2/
deltaDose^2 + (kki^2+kkj^2)/deltaDistance^2);
```

(basic expression – as implemented in the basic *myGammaSimple*)

```
norm ([(doseEvaluated (i, j) - doseReference (i+kki, j+kkj))/
deltaDose, kki/deltaDistance, kkj/deltaDistance]);
```

(Expression using *norm* – as implemented in the *myGammaSimple_norm*)

Local Normalization of Dose Difference

Difference in input: *doseEvaluated* and *doseReference* may or may not be normalized dose 2D-arrays as normalization of dose difference is performed separately for each gamma calculation. Nevertheless, for normalized input, the result will be the same as non-normalized only if the same normalization factor is applied to both 2D-arrays. Based on whether input 2D-arrays are normalized or as absolute doses in Gy, the *Threshold* parameter must be specified accordingly.

For the script *myGammaLocalSimple* presented in Web Supplement 9.1, local normalization by respective evaluated dose was chosen.

Simple Square Raster Scanning Search Space

For calculation efficacy purposes, one can consider replacing complex conditions of scanning in 'rings' around the evaluated point position by simply scanning the meaningful search space of the whole square (rectangle near borders). Such an approach can be demonstrated by the top two pictures in Figure 9.1: the whole meaningful search space determined by the Γ_I is addressed one at a time in a square raster manner and is similar to addressing original evaluated points in the basic algorithm.

Dynamically Adapted Search Space

With 'mathematically meaningful search space' defined by the first gamma Γ_I, one can consider dynamically adapting the search space using 'current best' (lowest) gamma for further principal search space reduction and a potentially faster calculation. The reasoning is exactly the same as explained in the basic Algorithm Scheme discussed earlier – that is, the current best gamma determines the maximum search space radius beyond which each gamma must be larger.

The script *myGammaSimpleDynSearchspace* is the original *myGammaSimple* script modified accordingly to compare both approaches. As dynamic adaptation of search space is a natural option the original *myGammaSimple* script is based on rather 'ring like' than 'raster like' scanning, making including the dynamic adaptation trivial.

Arbitrary Restricting Meaningful Search Space

Although from a mathematical perspective it makes sense to search the minimum Γ within the whole *k_radius* defined search space, it may not be so from a clinical perspective. If, for example, *k_radius* based on the Γ_I is large (eg, 10) then based on the search space defined so far, we need to compare 21 × 21 dose points which is computationally demanding considering this all relates to one evaluated point only. Also any Γ value on the search space periphery will be more than 10, far beyond pass/fail tolerance of 1. For this reason it may be preferable to accept potentially missing the true minimum Γ since from a clinical perspective it usually does not matter if the evaluated point fails with γ equal to, for example, 11.4 or 10.6. Even if a lower gamma is found within such a distance it is more likely to be a coincidence than a clinically meaningful reference point with the best match to a given evaluated point. For these reasons, it is useful to introduce additional speed up factor restricting 'clinically meaningful search space'. For dose 2D-arrays with fixed resolution of 1×1 mm^2, this can be 10 (points/millimeters) so the ultimate search space radius is *min {k_radius, 10}*. Such 'arbitrarily restricting search space' parameters may as well be introduced as the next user specified input for the gamma calculating function.

No Threshold Test for Each Evaluated Point

Instead of testing dose value against the *Threshold* dose for each point of the evaluated dose 2D-array (*doseEvaluated*), one can consider pre-selecting eligible points in one global test beforehand and calculating gamma only for these points without the *Threshold* test. In theory such an approach spares one operation for each evaluated point.

No Out-of-Border Test for Each Gamma Calculation

Following a similar approach as discussed previously, one can consider eliminating the out-of-border test for each gamma calculation. Square shaped 'meaningful search space' (see the basic Algorithm Scheme in the section 'Explanation of the Proposed MATLAB Scripts to Calculate the Gamma Index for Two Given Dose Distributions' earlier in this chapter) and Figure 9.1) can be defined immediately after the *k_radius* parameter is determined for a given evaluated point allowing addressing only those reference points meeting both criteria of being within the search space and not outside the reference 2D-array at the same time. Again, the major motivation is to replace two out-of-border conditions check for each gamma calculation by a single operation performed before the gamma loop has started.

Circular Search Space

So far, meaningful search space has been defined as a square around the evaluated point position determined by the first gamma Γ_1. From the definition and purpose of the meaningful search space it is obvious that this really should be a circle. So points making a square larger than a circle with the same diameter as the square side are included unnecessarily to gamma calculation since their distance to the evaluated point position is larger than the meaningful search space parameter. Therefore, one could consider adjusting the code that points from either whole reference 2D-array or square-shaped search space are tested for the Euclidean distance to the evaluated point position and only those points with distance smaller than meaningful search space parameter qualify for gamma calculation. This would result in fewer calculations of gamma in total for the cost of one extra condition for each reference point considered.

Basic Gamma Calculation in 3D

The three-dimensional version of the basic gamma algorithm assumes 3D input dose 2D-arrays with identical respective dimensions. The 2D-arrays are aligned with uniform spatial resolution of 1 mm^3 and normalized globally. It is the same for the 2D situation, the resulting gamma 2D-array have identical dimensions. Also, gamma statistics for the final result display is handled identically. So the only difference in fact, is 3D scanning instead of 2D scanning which is easily added by incorporating an extra dimension in both (evaluated points and search space) voxel addressing loops. Of course, calculating 3D distance requires adequate alteration as well. A graphical display of the result cannot be simply a gamma map this time, so one option is to use three figures instead of 1 for the 2D problem – one for each major anatomical direction (transverse, coronal, sagittal) – through the position of the global reference dose maximum as an arbitrary but reasonable choice.

Generalised Gamma Calculation Algorithm

Due to finite spatial resolution of the reference dose 2D-array, it is possible that for a given evaluated dose point the gamma corresponding to 2 neighbouring reference points from the search space (ie, with very similar or even identical geometry component) will be larger than 1, even for situations when the respective dose difference to the evaluated point dose is positive in one case and negative in another. This can happen in reference dose gradients that are steep relative to spatial resolution. If, at the same time, these 2 model gammas are smallest from the entire search space (and geometry component is near DTA tolerance), the overall test result for a given evaluated point will be negative. However, it is clear that the 'somewhere between these 2 points' dose difference, component is zero and the gamma is determined just by its geometry component and the γ for given evaluated point should be smaller, potentially below unity. The frequency of these events will increase with steepness of reference dose gradients and with dose pixel/voxel size. Therefore, if rather 'accurate' than 'reasonably accurate' gamma calculation is required, then interpolation should be incorporated into the algorithm. Of course, this will increase calculation time.

Algorithm Scheme: Interpolation Specific

Function Syntax

```
myGammaGeneral (doseEvaluated, spacingEval, refPointEval,
doseReference, spacingRef, refpointRef, deltaDose, deltaDistance,
Threshold)
```

Function Input/Output
  Input

doseEvaluated:	Evaluated 2D dose distribution, rectangular 2D-array, normalized globally
spacingEval:	Evaluated dose 2D-array (isotropic) spatial resolution in [mm]
refPointEval:	Evaluated dose 2D-array geometric reference point [row, column]
doseReference:	Reference 2D dose distribution, rectangular 2D-array, normalized globally
spacingRef:	Reference dose 2D-array (isotropic) spatial resolution in [mm]
refPointRef:	Reference dose 2D-array geometric reference point [row, column]
deltaDose:	Dose difference tolerance in [%]
deltaDistance:	Distance-to-agreement tolerance in [mm]
Threshold:	Threshold level for gamma calculation in [%] and *doseEvaluated*

Note: Geometric reference point forms a geometric link between the two dose 2D-arrays, for example, isocenter, central beam axis, or any other invariant specified for both dose distributions.

Output:

- 2D-array of the gamma values of the same size as *doseEvaluated* with negative ones at non-evaluated points to exclude them easily from the statistics
- Figure displaying the gamma 2D-array with dedicated gamma colormap: white for non-evaluated points, grayscale (white to black) for gamma from 0 to 1, and red for $\gamma > 1$
- On-screen display absolute number of test failing points
- On-screen display pass fraction of all evaluated points
- On-screen display number of evaluated points with interpolation involved

Preparations: General Algorithm Specific

- Convert the evaluated dose 2D-array into the list mode: each dose point is a 1D-array *[x, y, D, i, j]* where *x, y* are distances from the *refPointEval* in the respective dimension in mm, *D* is the dose and *i, j* are respective 2D-array indices
- Convert the reference dose 2D-array in the list mode: same as above for the reference dose 2D-array

Loop: Calculate the gamma for each evaluated dose point

- For each dose point from the evaluated dose list (*n*) ...
- Check the evaluated dose list *listDoseEval (n,3)* is larger than *Threshold* dose. Do nothing if not. If so then ...
- Find geometrically closest reference dose point from the reference dose list *listDoseRef*. The closest point has the smallest *x-y* distance (from the geometric reference point) difference between the evaluated and all reference dose points
- From the closest reference point data extract respective 2D-array indices *(i, j)* and proceed with calculating the first gamma Γ_1 to define the 'mathematically meaningful search space' (*k_radius*) as for the basic algorithm described earlier
- Note, searching the smallest gamma Γ is performed again in the (reference) 2D-array mode, ie, the list mode of expressing dose points is used only to address evaluated points and to find the closest point to handle distances *x* and *y* to define the search space center effectively
- Loop: Searching through the meaningful search space defined by the *k_radius*
  - Scan through the search space in the same way as for the basic algorithm (see Figure 9.1)
  - If the Γ is smaller than current smallest then replace the current smallest
    - Note the 'only' difference to the basic algorithm so far is implementing variable spatial resolution and 2D-array dimensions for dose distributions being

compared by using the list expression of dose points to address *x, y* distances and search for the closest point in the reference dose 2D-array efficiently ...

– Interpolation specific parts of the algorithm are coming now ...

• See Figure 9.2: First, test the sign of dose difference between a point in the search space (*kki, kkj*) and the evaluated point dose. If the sign is different to the respective sign of dose difference corresponding to the search space center (the closest to the evaluated point position in the reference 2D-array), then it is obvious that 'somewhere between' the search space center and the current reference point the dose difference is zero and interpolation should be involved ...

• This 'somewhere between' is determined as 1 or 2 points (depending how much off-center the subjected reference point (*kki, kkj*) is) with 2D-array indices lower by 1 in respective dimension(s), ie, based on whether *kki >/</= kkj* respective one or both search space 2D-array indices drop by 1 to determine these 1 or 2 points towards the search space center ...

• These 1 or 2 test points are tested for signs of dose difference again and in case this is different to the subjected (*kki, kkj*) point 1 or 2, interpolations are done between the subjected reference point and the test point(s) ...

– The interpolation consists in searching the distance to the evaluated point position (the search space center) for a position between the two reference points corresponding to zero dose difference

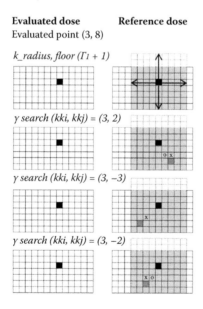

FIGURE 9.2 Demonstrating the definition of 'mathematically meaningful search space' for the minimum Γ (same as the basic algorithm) and examples of three reference points from the search space (*kki, kkj*) with interpolation included. When a reference point from the search space exhibits a different sign of dose difference compared with the reference point at the evaluated point position (the search space center), one or two test points (X, O) in direction towards the center are used for interpolation in order to provide additional one or two gammas (Γ) with geometry only component. The smallest Γ always applies.

- Each calculated gamma(s) Γ is compared with the current smallest, replacing it when smaller
- The search space parameter, *k_radius*, is adjusted for each new (smaller) gamma (see the section 'Explanation of the Proposed MATLAB Scripts to Calculate the Gamma Index for Two Given Dose Distributions/Alternatives to the Basic Gamma Calculation Algorithm/*Dynamically Adapted Search Space*' earlier in this chapter)
- Carry on the loop for each dose point from the evaluated dose list (*n*) ...

Statistics and results display are equivalent to the basic algorithm described earlier.

Pass/Fail Only Gamma Calculation

It depends on the user's preference how accurate a gamma calculation is required for a given application. If 'only' a pass/fail fraction is satisfactory, it is possible to terminate searching for the minimum Γ at any time when the current minimum gamma falls below 1, allowing the given evaluated point to pass the given tolerances.

Of course in terms of result presentation, only pass/fail statistics and binary pass/fail gamma map for graphical output apply, since from the application perspective, it is pointless to present terminated gamma values.

Be aware that terminating search when gamma drops below 1 has no effect on overall pass/fail fraction, that is, the basic algorithm *myGammaSimple* has the same pass/fail fraction as 'termination included' *myGammaPassFailOnly*. It is other algorithm modifications, such as introducing the interpolation (see the section 'The Gamma Index for Two Given Dose Distributions/Alternatives to the Basic Gamma Calculation Algorithm' earlier in this chapter), for what affects the pass/fail fraction.

DEMONSTRATING MATLAB SCRIPTS TO CALCULATE THE GAMMA INDEX FOR TWO GIVEN DOSE DISTRIBUTIONS

Let us assume there are two 2D/3D dose distributions available in MATLAB workspace meeting the following criteria:

- Both evaluated and reference dose distributions have identical spatial resolution 1×1 mm^2/$1 \times 1 \times 1$ mm^3 for 2D/3D case, respectively

- Both dose 2D-arrays have the same size and are aligned

- Both dose 2D-arrays are normalized for both global and local normalization approach demonstration, or not normalized, ie, absolute dose in Gy, for local normalization approach demonstration

For an introduction to dose distributions import in MATLAB, see Chapter 3.

If any 2D-array has other than 1×1 mm^2/$1 \times 1 \times 1$ mm^3 resolution, then see Chapter 2 for change spatial resolution/resampling.

If 2D-arrays are not aligned then see the section on matching 2D data in Chapter 2.

If any 2D-array is not normalized and it is required, then see the section 'Dose Normalization' earlier in this chapter.

Sample dose 2D-arrays meeting all of the above criteria are available in Web Supplement 9.2.

- *ref_1x1_aligned_Gy*

- *eval_1x1_aligned_Gy*

Three-dimensional dose examples:

- *ref_1x1x1_aligned_Gy*

- *eval_1x1x1_aligned_Gy*

MATLAB syntax demonstrating basic gamma calculation script example and alternatives are presented in this chapter. Scripts working with identical 2D sample dose data and parameter values are presented first, one 3D script for 3D sample dose data follows later. Using fixed sample data and parameter values enables efficient demonstration of individual algorithm/script specifics.

Basic Algorithm and Its Alternatives – 2D
Demo syntax:

```
>> load gammaColormap
>> load ('eval_1x1_aligned_NormRefDmax.mat');
>> load ('ref_1x1_aligned_NormRefDmax.mat')
```

… shorten the variables names

```
>> doseRef = ref_1x1_aligned_NormRefDmax;
>> doseEval = eval_1x1_aligned_NormRefDmax;
```

… and calculate the gamma maps with various algorithm versions described above using scripts available in Web Supplement 9.1

```
>> gammaSimple = myGammaSimple (doseEval, doseRef, 3, 3, 10);
>> gammaSimpleReversed = myGammaSimple (doseRef, doseEval,3,3,10);
>> gammaSimple_norm = myGammalSimple_norm (doseEval, doseRef, 3, 3,
   10);
>> gammaLocal = myGammaLocalSimple(doseEval, doseRef, 3, 3, 10);
>> gammaSquareRaster = myGammaSimpleSquareRaster (doseEval, doseRef,
   3, 3, 10);
>> gammaDynSearchspace = myGammaSimpleDynSearchspace (doseEval,
   doseRef, 3, 3, 10);
```

```
>> gammaArbRestrictSearchspace = myGammaSimpleArbRestrictSearchspace
   (doseEval, doseRef, 3, 3, 10);
>> gammaPreThreshold = myGammaSimplePreThreshold (doseEval,
   doseRef, 3, 3, 10);
>> gammaPreThresholdSquareRaster = myGammaSimplePreThreshold
   SquareRaster (doseEval, doseRef, 3, 3, 10);
>> gammaPreThresholdPreBorder = myGammaSimplePreThresholdPreBorder
   (doseEval, doseRef, 3, 3, 10);
>> gammaPreThresholdCircSearch = myGammaSimplePreThresholdCircSearch
   (doseEval, doseRef, 3, 3, 10);
>> gammaGeneral = myGammaGeneral (doseEval, 1, [150, 150],
   doseRef, 1, [150, 150], 3, 3, 10);
```

Basic Algorithm – 3D:

Demo syntax:

```
>> load gammaColormap
>> load ('eval_1x1x1_aligned_Gy'); load ('ref_1x1x1_aligned_Gy')
```

… normalize, shorten the variables names and test the script

```
>> doseEval = eval_1x1x1_aligned_Gy/max (ref_1x1x1_aligned_Gy (:))
   * 100;
>> doseRef = ref_1x1_aligned_Gy/max (ref_1x1x1_aligned_Gy (:)) *
   100;
>> gamma3D = myGammaSimple3D (doseEval, doseRef, 3, 3, 10);
```

Simple comparison of all 2D algorithm versions tested in terms of quantitative result and calculation demands is presented in Table 9.1. Looking at the figures one can observe few outcomes:

- Raster scanning of the search space seems slightly quicker compared with concentric 'circles' with increasing radius

- Using the intrinsic *norm* function for gamma calculation seems significantly slower compared with direct gamma calculation formula

- Using a dynamic adaptation of the search space seems quicker compared with 'first gamma only based' search space

- Using arbitrary restricted search space (clinically meaningful search space) has an impact on calculation speed compared with gamma-only based search space, however, in principle, this approach affects the overall result so is not absolutely equivalent

- Pre-selecting eligible evaluated points seems slightly slower compared with the testing dose *Threshold* condition for each point in the evaluated dose 2D-array

- Cropping search space by reference dose 2D-array borders seems significantly slower compared with performing out-of-border test for each tested reference point

TABLE 9.1 Comparing various gamma calculation algorithm versions in terms of quantitative results and calculation demand

Algorithm Version		Points Failed	Points Evaluated	Pass Fraction [%]	Calculation Time [ms] *)
2D	*myGammaSimple*				360
	_norm				3050
	SquareRaster				282
	DynSearchspace				260
	ArbRestrictSearchspace	1149	52354	97.80	332
	PreThreshold				384
	PreThresholdPreborder				5910
	PreThresholdCircSearch				3280
	myGammaGeneral	1031	52354	98.03	250615
	myGammaLocalSimple	5229	52354	90.01	310
	evaluated<->reference	14712	65996	77.70	9730
3D	*myGammaSimple3D*	75392	950399	92.06	24680

Note: Single basic 3D test and the importance of a choice between evaluated and reference dose are included as well.

*) Best of 10 runs on Intel i5/MATLAB R2013a/MS Windows 7.

- Including 'radius-based' condition to shrink the search space is significantly slower compared with scanning entire 'square-like' search space

- Using local normalization of dose difference is comparable with global normalization standard in terms of calculating demands

- Including interpolation in the algorithm is significantly more demanding computationally but produces more accurate overall gamma result ie, larger pass fraction eliminating the majority of false negatives due to gradient-resolution aspect typical for simple algorithm

- Using local normalization of dose difference results in significantly smaller pass fraction (larger number of failing points) compared with the global normalization standard

- Significant difference in both calculation demands and overall quantitative result when the evaluated and the reference doses replace one another in otherwise identical gamma calculation

- Compared with 2D, the 3D gamma calculation is generally more demanding due to both larger number of evaluated points and larger search space

Three different gamma maps of all 2D gamma scripts tested are shown in Figure 9.3. In agreement with expectations, compared with the basic globally normalized gamma algorithm without interpolation, the red areas representing failing gamma points are relatively larger for the locally normalized dose difference, and relatively smaller for interpolation including algorithms. Additionally, the figure shows the gamma map for the basic algorithm and identical

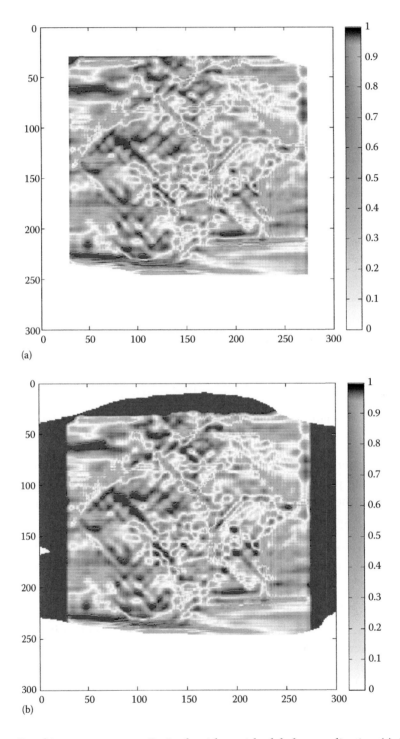

FIGURE 9.3 Resulting gamma maps: Basic algorithm with global normalization (a), basic algorithm for reverse order of the evaluated and reference sample doses (b). *(Continued)*

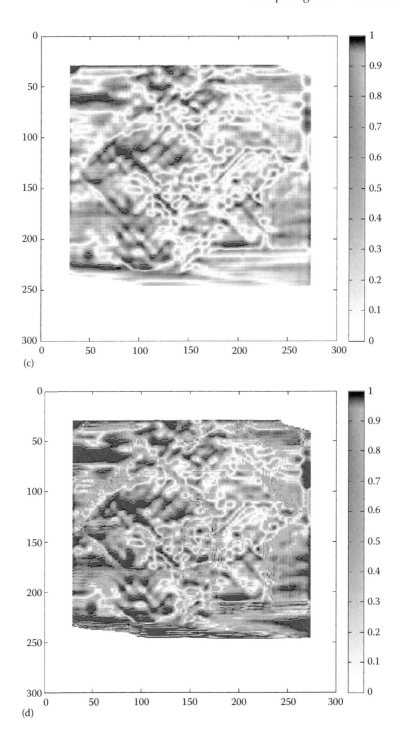

FIGURE 9.3 (CONTINUED) Resulting gamma maps: General algorithm with global normalization of dose difference (c), and basic algorithm with local normalization (d). Pass fraction for 10% *Threshold* and same conditions otherwise is 97.8%, 77.7%, 98.0% and 90.0%, respectively.

input parameters otherwise for the evaluated and reference dose distributions reversed. The results clearly demonstrate a significant difference in the quantitative result, essentially due to differences in dose threshold for evaluation criterion, and also nature of dose distributions: the dose labelled as the 'evaluated dose' used for this example was a measured dose while the 'reference dose' was calculated by a treatment planning system. In general, calculated dose distributions are smoother than measured ones with significant impact on computing demands for the search space scanning when looking for the minimum gamma.

It is important to understand that the presented outcomes and even basic algorithm alternatives do not represent an optimally coded gamma algorithm. Rather, they demonstrate aspects and parameters with either a smaller or larger effect on calculation accuracy and calculation time.

FURTHER PROCESSING GAMMA MAPS

Although basic statistics resulting in the 'pass fraction' certainly represent the most common standard in result presentation, some may prefer alternative or extended approaches. Several examples are presented in the next section.

Mean Gamma Value of All Evaluated Dose Points

The pass/fail fraction tells exactly that fraction of evaluated points did or did not meet $\Delta D|\Delta d$ tolerances. It does not say anything about 'by how much' the points failed or passed. It also doesn't say anything about 'whether the reference dose is larger or smaller than the dose being evaluated'. For these, and possibly more reasons, some prefer using metrics that include the gamma result of all evaluated points. One of the easiest is the mean gamma.

From the gamma 2D-array function outputs, one can obtain such statistics easily (for the basic *gammaSimple/myGammaSimple* example):

```
>>  mean (gammaSimple (gammaSimple > 0))
```

- Calculating the mean value of the gamma 2D-array elements with 2D-array indices corresponding to non-negative gamma value (ie, the evaluated points with dose above *Threshold*)

- Works for both 2D and 3D arrays

Gamma Histogram as a General Extension of Pass Fraction Statistics

The histogram shows more information about the distribution of the range of 2D-array elements values compared with a single parameter such as pass fraction ($P_{\gamma<1}$). The gamma histogram can be obtained easily in MATLAB and the only question that needs to be addressed is what histogram bins would provide the required information, for example,

```
>> gammaHist = histc(gammaSimple(:), [0:.1:2, inf])
>> figure; stairs([0:.1:2,2.1], gammaHist)
```

- Specified bin 1D-array values together with the step of 0.1 and choice of the *histc* in-built function provide direct information about $P_{\gamma>1.5}$ or $P_{\gamma>2}$

- An arbitrary maximum value of *2.1* is selected for correct display of the last bin corresponding to $2 <= \gamma < inf$

- *Stairs* rather than the *bar* plot command is preferred for a straightforward display of the histogram values vs. bin edges

- Works for both 2D and 3D arrays

Inverse Alternative to Gamma Histogram

Standard γ-based dose comparison deals with a problem: 'What is the fraction of the ROI passing preset ΔD and Δd tolerances'? In some situations, however, it might be more useful to deal with the rather inverse problem: 'What are the minimum tolerances ΔD and Δd such that (*x*%) of the ROI (*x%ROI*) pass the test'?

For example, in common practice users calculate the gamma for standard 3%|3mm tolerances. When more points than the pre-agreed tolerance fail, then either further investigation without using gamma is performed, or, more commonly, the next gamma is calculated for larger tolerances, 4%|3 mm and the result assessed. The point is that in practice it is not rare when more than one gamma calculation is required. In addition, users naturally tend to choose integer and rather symmetrical tolerance values so 3.7%|3.4 mm tolerances are not intrinsically included in the assessment, although it could be these values which determine the boundary when the pass fraction crosses from 'unacceptable' to 'acceptable'. Another aspect is that from a mathematical perspective the ratio $\Delta D/\Delta d$ represents relative weight or importance of respective quantity. Thinking deeper about the relative importance of the two quantities, from a rather clinical perspective, one probably would give relatively higher importance to geometrical rather than dose parameter, mainly for a relatively larger uncertainty in dose-effect relationship (difference in effect for dose difference of 2% vs. impact of an error in dose location of 2 mm?). These are reasons for having the availability of a method providing multiple continuous $\Delta D|\Delta d$ tolerances for a fixed preset passing fraction values (*x%ROI*) per unit gamma calculation as a useful alternative to direct approach.

The method presented in Dvorak (2009) enables one to consider pass/fail criteria or action levels in terms of a fixed passing fraction *x%ROI* and restricted (maximum allowed) minimum $\Delta Dx\%ROI \mid \Delta dx\%ROI$ tolerances, instead of standard restricted (maximum allowed) *x%ROI* for fixed $\Delta D|\Delta d$ tolerances.

The essential prerequisite of the method is the fact that the ΔD and Δd represent mathematically normalization constants, and that the gamma value of two dose points being compared with a given multiple of the original $\Delta D|\Delta d$ tolerances is equal to the original gamma divided by this multiplication factor. For example, if the gamma for 2%|3 mm tolerances is equal to 1 then the gamma of the same two points for 4%|6 mm tolerances is equal to 0.5, that is, smaller by the tolerance multiplication factor 2.

This aspect of the gamma has potential in testing, particularly the algorithm for consistency of the gamma calculation. In algorithms with this rule corrupted often the suspect is some simplification included to speed up the calculation process. Uncovering search space over-reductions may be a good example of an application of such a simple test. See Dvorak (2009) for more details about how to implement the alternative to histogram presentation of the gamma result.

MERITS AND LIMITATIONS OF GAMMA ANALYSIS

The gamma method represents the contemporary standard for quantitative comparison of dose distributions in radiotherapy physics. All commercial software comparing dose distributions includes some form of the gamma algorithm. Some products offer even more options. Algorithm versions available may and should differ in the essential choice of normalization of dose difference (global vs. local). Some products may require normalization before gamma calculation and the user should be fully familiar with the options and methods available. Some products may offer two algorithm versions, both utilizing global normalization, with one version resulting in a standard outcome with a full range of gamma values, and another with a binary pass/fail only outcome. The latter version may offer 'only' a pass/fail result, but this result may be more accurate than the alternative, full gamma, algorithm simply because developers decided to compromise accuracy (eg, not including interpolation) in the 'full gamma' version. Which version to use is clearly the user's preference depending on the objectives.

Result Interpretation

Thinking about the whole range, from simple dose difference based dose comparison to the gamma method with the dose difference tolerance *and* distance-to-agreement tolerance *and* a tolerance for resulting pass fraction (or tolerances applied to the gamma histogram derived parameters) for all other parameters involved, it is obvious that instead of a single decision level there are now two ($\Delta D | \Delta d$), three (+*Threshold*) or four (+acceptable *pass fraction*, mean gamma, etc.). This makes result interpretation more complicated mostly because of the inclusion of the geometric factor. The gamma combines dose difference and DTA which is an advantage in high dose gradients, so typical for modern techniques, preventing large dose differences due to resolution or a small geometrical mismatch. The same might be seen as a disadvantage from losing the added information in the two distinct views. A question that probably describes the problem well is 'Is it really OK if the prescribed dose is, for example, 3 mm away from what is desired'? Obviously, the answer depends on application and, to some extent, on the user's subjective assessment.

In addition to the gamma map there is one complementary parameter which may help in interpreting the gamma result. This is the *gamma angle*, that is, the angle of the gamma vector with respective dose difference and distance components in the dose-distance space calculated as a

$$atan(r / \Delta d / \delta * \Delta D) \tag{9.6}$$

The gamma angle range is from 0 to 90° corresponding to two extreme situations:

- The smallest Γ was found for the dose difference component only at the identical reference point location as the evaluated point (gamma angle = 0°)

- The smallest Γ was found for DTA only component (gamma angle = 90°)

   That is, a simultaneous assessment of both the gamma and the gamma angle maps may help distinguish whether points fail rather due to dose or geometric mismatch.

Practical demonstration of this approach is available using the *myGammaSimpleAngle* script available in Web Supplement 9.1. For more details about the gamma angle concept see Stock et al. (2005).

Input Sensitivity and Output Robustness

In terms of sensitivity and robustness, it is useful to consider clinical requirements first. Although there are many parameters determining or influencing the gamma result, let us focus on the basic tolerances $\Delta D|\Delta d$ and character of the dose distribution alone. From a clinical perspective it is required that the method is sensitive enough to detect even a minor alteration to the desired dose. On the other hand, it may not be so beneficial if the method is oversensitive to a choice of more or less arbitrary tolerances such as most common 3%|3 mm. From a clinical practice we know that gamma is quite sensitive to $\Delta D|\Delta d$, which is also why when the pass fraction is 'too low' it is common to increase tolerances slightly and hope for a 'good enough' pass fraction in order to pronounce the test successful. For example, the two samples from Web Supplement 9.2 show a pass fraction for respective ΔD tolerance in the range from 94% (2%) to 99% (4%) (same parameters otherwise). However, each dose distribution and each comparison test is different and the magnitude and range of possible situations is large so it is difficult to attempt for some general rules. Obviously, all gamma-based statistics will depend strongly on *Threshold* or any other alternative definition of region of interest.

One can only recommend application specific parameters and methods including actions when using the gamma method for a quick quantitative comparison of two dose distributions in clinical practice.

Specification of a Particular Gamma Assessment

Unfortunately, details about a particular gamma implementation in application software helps and manuals are sometimes insufficient which is not surprising considering the complexity and large number of parameters involved.

One of the major motivation factors to including a chapter about the gamma calculation was the established standard of reporting results. A common reporting standard is a pass fraction for $\Delta D|\Delta d$ tolerances given in [%] and [mm]. However, as demonstrated in this chapter, this is far from a full description of what has been actually calculated and how. As we know, quantitative results of comparing two distributions using the gamma method depends further on:

- Which distribution was used as evaluated and which as reference

- What was the evaluated area/volume (eg, defined by evaluated dose threshold)

- How was dose difference normalized (globally to a single point [which one? or locally] – evaluated or reference used for difference normalization?), etc.

- Spatial resolution of reference dose distribution

- Limited spatial resolution handling

- Methods of restricting search space

Although from a scientific or mathematical perspective these arguments are certainly relevant, the gamma method was proposed for clinical applications. And from a clinical perspective it is certainly less dramatic when the pass fraction is 97.8% instead of 98.2% or, when the given large fraction of points do not meet 3%|3 mm tolerances but 3.2%|3.4 mm they do, and so on. For this reason, even a relatively less accurate algorithm can still serve its purpose. On the other hand, applications of the gamma method in clinical audits or inter-departmental or inter-device comparisons may be relatively more exposed to the effects related to the complexity of the precise method specification.

Nevertheless, the local implementation for a given application (eg, pre-treatment plan verification) or given device (eg, detector with associated control and analyzing software) is certainly achievable provided all aspects mentioned in this chapter have been considered during the definition of the appropriate standard. Careful definition of operational parameters, together with pass/fail criterion and appropriate action(s) is required for meaningful and safe application of the gamma method.

Clinical Relevance of the Gamma Result

If we want to discuss clinical relevance we need a relevant assessment metrics. An established standard of simplified expression of dose distribution quality is the dose volume histogram (DVH) relating volume fraction of a volume of interest (VOI) and the minimum dose this fraction receives (see the section 'DVH: Differential and Integral Format' in Chapter 2). Notwithstanding the loss of dose distribution information arising from its simplicity, most organ specific radiation toxicity parameters shared within the community today are based on DVH making this metric acceptable for the purposes of this chapter. Considering the clinical relevance of the gamma method, we can ask whether there is a positive dose comparison result in terms of the gamma method with clinically unacceptable related DVH parameters, and vice versa. It is probably not surprising that dose distribution comparisons with both situations described do exist, especially considering the most conventional application of the gamma method with the pass/fail fraction as the quantitative outcome. The reasons are obvious: simple statistical parameters do not distinguish the location of a particular disagreement, the gamma does not distinguish negative or positive dose difference, and the Δd (DTA) tolerance may simply be too large to detect a big change of dose close to a given structure border causing potentially significant difference in a particular DVH parameter. More about the gamma and DVH parameters can be found in Jin et al. (2015).

In order to distinguish location and under-/over-dose situations, the gamma maps and/or related statistics can be dealt with separately for points where the reference dose is larger or smaller than the evaluated dose, respectively, effectively forming further ROIs. Similarly, clinical VOIs, usually available with (3D) clinical dose distributions or their projections to respective radiation field direction (2D), can be used as 'clinically relevant ROIs' enabling clinical structure-specific gamma statistics. Using these and similar tools the use gains more specific information about dose distribution mismatch in potentially clinically relevant regions. However, sometimes qualitative, that is, visual, assessment of the gamma map possibly with appropriate graphical interpretation cannot be replaced by any quantitative statistics. When using VOIs or their projections, a user should be aware of the principle difference between dose distribution that results from a given treatment plan on a patient model and on a phantom potential disagreement within volume/area of a given structure as contoured on patient CT model. This is mathematically transferred on a phantom to perform a plan verification measurement, and is not exactly interpretable as identical disagreement for the given structure on a patient CT model. Clearly, this is not a problem for dose assessments where both evaluated and reference doses originate from the patient model.

Alternative Methods

As mentioned earlier, DVH is the current standard clinical indicator of a given dose distribution quality. From this perspective one would expect comparing two sets of DVHs and/or derived statistics (eg, VOI mean dose, $V_{95\%}$, V_{20Gy}, D_{max}) for relevant clinical structures as the ultimate method of comparing dose distributions. Given examples of DVH-derived parameters can be further extended to potentially more clinically relevant radiobiological parameters such as tumor control probability (TCP) and normal tissue complication probability (NTCP). One problem is that DVH-based assessment is possible only for 3D doses so it is available only for 3D dosimetric systems or alternatively 3D dose calculations. Another problem is that clinically it may be difficult to say at which point the difference of a particular DVH-based parameter becomes clinically significant. Therefore, whatever pass-fail criterion, whether based on the dose difference, the gamma statistics, or DVH-based parameters, is, to some extent, subjective even though the major motivation of quantitative dose comparison was objectivity of the assessment. Basically, even clinical parameters such as those based on DVH are not specified with accuracy high enough to distinguish clearly 'acceptable range' vs. 'non-acceptable range' of parameter values. This reflects the principle differences between 'rather medicine' and 'rather physics' approaches in the sense that a clinical decision is not and cannot follow the objective 'accurate' standards of physics for many reasons, one of which is ethics: it would be unethical to test all possible options on patients to gather sufficient and robust data to derive strong rules about an acceptable range of a particular parameter. In addition, patient responses to identical parameters may vary.

For 2D dose distributions there is no substitute to the DVH. Hence, the clinical relevance of dose comparison methods remains rather indirect and more linked to a physics perspective. As mentioned in the beginning of this chapter, comparing 2D dose distributions is common in linac and TPS commissioning or QA. Some objectives may include complex dose distributions, for example, testing MLC performance and/or MLC parameters using a dedicated IMRT plan. There are also rather simple 'very little modulated' radiation fields and the resulting 2D doses such as open and/or wedged fields of various shapes. For such situations the gamma method does not make much sense owing to the irrelevance of the

DTA component of the gamma vector and alternative metrics such as simple dose difference or even qualitative assessment may be more suitable. High dose gradients areas are typically only along field edges which is easily detectable, explainable and correctable. Otherwise, for complex dose distributions, one can consider probably many methods or approaches with their pros and cons.

REFERENCES

De Martin E, Fiorino C, Broggi S, Longobardi B, Pierelli A, Perna L, Cattaneo GM, Calandrino R. (2007). Agreement criteria between expected and measured field fluences in IMRT of head and neck cancer: The importance and use of the gamma histograms statistical analysis. *Radiotherapy and Oncology* 85(3), 399–406.

Dvorak P. (2009). An alternative to gamma histograms for ROI-based quantitative dose comparisons. *Physics in Medicine and Biology* 54, N247.

ESTRO. (2008). *ESTRO Booklet No. 9: Guidelines for the verification of IMRT.* Brussels, Belgium: ESTRO.

IPEM. *Guidance for the clinical implementation of intensity modulated radiation therapy.* York, United Kingdom: IPEM.

Jin X, Yan H, Han C, Zhou Y, Yi J, Xie C. (2015). Correlation between gamma index passing rate and clinical dosimetric difference for pre-treatment 2D and 3D volumetric modulated arc therapy dosimetric verification. *British Jouurnal of Radiology* 88(1047), 20140577.

Low DA, Harms WB, Mutic S, Purdy JA. (1998). Complex A technique for the quantitative evaluation of dose distributions one can consider probably many methods or approaches with their pros and cons. *Medical Physics* 25, 656–661.

Stock M, Kroupa B, Georg D. (2005). Interpretation and evaluation of the gamma index and the gamma index angle for the verification of IMRT hybrid plans. *Physics and Medical Biology* 50(3), 399–411.

Venselaar J, Welleweerd H, Mijnheer B. (2001). Tolerances for the accuracy of photon beam dose calculations of treatment planning systems. *Radiotherapy and Oncology* 60(2), 191–201.

CHAPTER 10

Example of Accessory Modeling in Radiotherapy

INTRODUCTION

This chapter presents an example of accessory (treatment couch) modeling in radiotherapy using a three-dimensional model of the treatment couch applicable to dosimetry in vivo using point detectors. The model consists of multiple analytical planes approximating the couch surface and allows the calculations of the two intersection points (entrance and exit) of the couch surface and a given beam rayline. In clinical practice, this information can be used to determine the source detector distance as well as the detector incidence angle for the in vivo dosimetry of posterior fields when the detector needs to be placed on the couch underneath the patient instead of on the patient surface. One can also use the model to estimate backscatter for accurate correction of the detector reading.

APPLICATIONS OF ACCESSORY MODELING

The radiotherapy process offers many examples of computer modeling. Planning CT series provides a 3D model of the patient to be used for treatment planning and for calculating the absorbed dose distributions. The core of each treatment planning system (TPS) is a computerized model of every treatment beam available on the given treatment machine. Modern TPSs act as *virtual simulators*, simulating or modeling beam geometry together with the given patient CT model in order to provide visual feedback to a treatment planner to guide optimization of beam geometry. This uses features such as beam's eye view (BEV) displaying a field shape and projected volumes of interest (VOIs) under the given simulated beam direction.

When speaking about a beam model, this usually consists of beam parameters required by a given dose calculation algorithm to determine a given beam dose contribution to a given point (voxel) in the patient body. As demonstrated in Chapter 7, beam data forming a base for a given beam model may look like a collection of dosimetric parameters rather than a mathematical description of a given radiation source, including any accessories such as beam collimation system, and so on. However, for Monte Carlo based dose calculation algorithms, this is required for its explicit particle transport modeling nature. The history of each generated photon is tracked on its way from the source (X-ray target) through the collimation system until it leaves the machine where it contributes to a convenient quantity (energy fluence map), which can be used further as the radiation source for a given patient model below to calculate local deposition of energy – the dose (AAPM TG 105 or Jabbari 2011).

Outside of the beam modeling area, there are two natural and simple example applications of accessory modeling in the geometrical sense that are considered in this chapter.

Collision Avoidance

When the operating limits of two or more components of the given treatment system allow potential collision, it is easily preventable by excluding colliding position combinations. These may be determined either through direct measurements, or by appropriate geometric modeling of the relevant components; both approaches leading to the identification of any colliding positions. Of course, this is much easier done for fixed machine components such as a linac gantry and treatment couch, than for 'moving accessories' such as immobilization devices and even the patient. To handle this type of collisions, modern machines are equipped with 'collision guard' systems based on optical principles (or similar).

Treatment Couch Attenuation

Conventional conformal radiotherapy often includes posterior (assuming patient supine position) beam directions which are attenuated through the treatment couch before entering the patient's body. For accurate radiotherapy this attenuation should be either minimized or accounted for in the dose calculation. This is why modern treatment couches take the form of a shell made of carbon fibers. Such construction provides the required mechanical performance with minimum beam attenuation so that when couch attenuation is ignored, the dose calculation inaccuracy is small, for example, 2%–3%, and decreases with the increasing number of beams. Alternatively, when the correct treatment couch is not present in the given planning CT series (which is typical) and also for situations when neglecting the carbon fiber couch absorption albeit low is still not acceptable, treatment planning systems must include appropriate treatment couch models. For more details on general attenuation aspects of couch tops and accessory see AAPM TG 176. For example, TPS Eclipse™ (Varian Medical Systems) offers three options to consider in their Exact® IGRT couch top dose correction. The couch top thickness options vary as the couch top thickness varies along its length ('thin', 'medium' and 'thick'). Based on the given anatomical site, patient orientation and estimated position on the couch, a treatment planner can choose the most suitable option. This chapter will also show that this particular couch top can be modeled more accurately in spite of its relatively complex shape. On the other hand, due to the couch shell-like construction, beam absorption depends little on the actual couch thickness so a single universal option provides an acceptable solution for most, if not all, clinical situations (Vaneti 2009).

DOSIMETRY IN VIVO

In vivo dosimetry in radiotherapy has been established as an independent method of verifying the correct dose delivery directly on a patient. This is recommended in many countries and is even mandatory in some. Traditionally, it has been based on measuring the entrance and/or exit dose for a single treatment beam by using a small point detector (typically a semiconductor diode) placed on the patient's skin. The given entrance and/or exit doses are compared with the expected values calculated by the TPS. For beams entering the patient's body through the couch, the detector has to be placed on the couch surface. In such a case the given detector conditions deviate significantly from the standard calibration conditions and its reading should be corrected accordingly. Standard calibration conditions for the entrance dose measurement often include reference parameters such as (open) field size, source surface distance (SSD), normal angle of incidence (AOI), defined backscatter, temperature, dose rate in terms of $MU.min^{-1}$, beam nominal energy and possibly more. The amount of backscatter during calibration is usually large enough to provide full backscatter conditions for typical measurement on patients. The AOI correction is required for detectors whose response is dependent on it. The three named parameters (SSD, backscatter, AOI) are relevant for situations when a given detector is placed on the treatment couch surface with intent to measure the entrance dose in a given patient body. The detector is not placed on the patient skin which is expected as per calibration conditions so:

- The beam is attenuated by the couch which may or may not be included in the calculated reference entrance dose. Based on that, the appropriate correction should or should not be applied

- The detector's response is higher than it would be at the patient's surface due to the shorter distance to the source. The source detector distance (*SDD*) does not equal SSD

- When the couch composition is not tissue equivalent, there are different backscatter conditions than considered at calibration

- The detector temperature remains rather ambient since it is less affected by the proximity of the patient's body as when placed on the skin

An appropriate 3D model of a given treatment couch can be considered when addressing the aspects listed (except temperature) in order to improve the accuracy of in vivo dosimetry using point detectors. Details of the practical aspects of implementing in vivo dosimetry using diodes can be found in ESTRO (2001).

TREATMENT COUCH MODEL EXAMPLE

The Exact IGRT couch top represents a widely used contemporary standard, therefore it is useful as a practical example of accessory modeling. The couch top is relatively complex in shape (see Figure 10.1) with variable thickness and rounded edges which all make a pretreatment prediction of *SDD* difficult. For detectors positioned on the beam central axis (CAX), it is easy to measure the actual *SDD* by using the linac's optical distance indicator (ODI) before a detector is attached to the couch. However, this is an awkward practice and it is often impossible for detectors placed off-axis which is typical for asymmetric fields. Furthermore, it is not always possible to accurately predict a patient's position on the couch or even the position of the couch before the first actual patient setup on a linac, notwithstanding the availability of an accessory indexing system.

First, the actual couch model was created by sampling the couch cross-section at five cross-section planes A, B, C, D and E. These five planes account for the couch shape in the longitudinal direction. Each of the planes contains 8 points in order to determine the couch shape in the respective cross section. The couch model origin was defined at the couch upper surface on the longitudinal couch axis at the '0' position of the couch longitudinal index. Coordinates of all reference points were determined by measurements with respect to the couch origin by both linac's ODI and a mechanical pointer. Having 40 reference points sampling the couch surface, the couch model alone

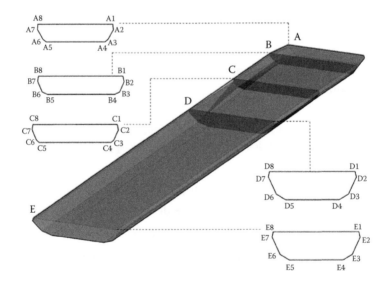

FIGURE 10.1 Example of the couch top model. Five sample planes (A, B, C, D and E), each consisting of 8 points and providing 40 couch model definition points in total. Neighbouring points form 43 couch model definition planes. These planes approximate the couch surface.

is formed by the system of 43 planes; each formed either by 4 neighbouring points when all 4 points are within the same plane such as: E1-E2-D1-D2, D1-D2-C1-C2, and so on, or 3 neighbouring points otherwise such as, for example: D2-D3-C2, D3-C3-C2, and so on (see Figure 10.1). The normal vector of each plane [a, b, c] is given by the cross product of any two independent directional vectors derived from the respective 4(3) points. Intercept d is then given by substitution for any of 4(3) points so that the analytical equation of a given plane is met:

$$ax + by + cz + d = 0 \qquad (10.1)$$

In the next step, the couch model has to be related to the actual treatment plan coordination system with the origin coinciding with the given treatment plan origin. This is achieved by recording the default couch position $[VRT_0, LNG_0, LAT_0]$ on each individual linac with the couch origin placed at the linac's isocenter. The couch model definition point coordinates with respect to the treatment plan origin will be then (for purposes of this demonstration let's consider the IEC 61217 treatment couch convention):

$$\begin{pmatrix} x \\ y \\ z \end{pmatrix} = \begin{pmatrix} x_{CouchOrig} \\ y_{CouchOrig} \\ z_{CouchOrig} \end{pmatrix} + \begin{pmatrix} x_{ISO} \\ y_{ISO} \\ z_{ISO} \end{pmatrix} + \begin{pmatrix} LAT - LAT_0 \\ LNG - LNG_0 \\ VRT - VRT_0 \end{pmatrix} \qquad (10.2)$$

where the index *CouchOrig* stands for the original measured XYZ coordinates relative to the couch origin.

The model can be then tested by by comparing the model source to couch distance with the distance measured using ODI and a mechanical pointer for various combinations of gantry and couch including the most curved surfaces. The accuracy of the presented simple model is to within 3 mm, which is equivalent to less than 1% dosimetrically in terms of inverse-square law.

CLINICAL APPLICATION OF THE MODEL

The couch top model described in the previous section can be easily implemented as a function script in MATLAB in order to determine the applicable parameters in dosimetry in vivo of posterior fields by using point detectors which are difficult to obtain otherwise. To explain the algorithm, consider the problem demonstrated graphically in Figure 10.2.

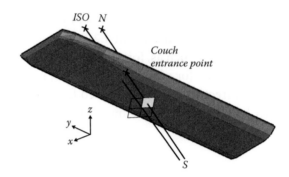

FIGURE 10.2 Let's consider a posterior quarter beam with off-axis in vivo dosimetry reference point N. The task is to measure the beam entrance dose 'above' the reference point N, ie, the point at reference depth (1.5 cm) below the patient surface on the source S–N rayline. It is assumed that the point detector is placed on the couch under the light field guidance. Three parameters are required for adequate detector response correction: source detector distance (SDD), angle of incidence (AOI) and the S-N rayline couch path length (couchPathLength).

Algorithm Scheme

Function Syntax
  couchIntersect (GNT, RTN, VRT, LNG, LAT, ISO, N)
Function Input/Output

  Input
    GNT, RTN   Linac gantry and couch isocentric rotation angle
    VRT, LNG, LAT   Couch vertical, longitudinal and lateral coordinate
    ISO   Linac isocenter XYZ coordinates from a given treatment plan
    N   In vivo dosimetry reference point XYZ coordinates from a given
                      treatment plan determining the rayline
  Output
    SDD   Source to detector (couch surface) distance
    couchPathLength   Distance between 2 intersection points (entrance and exit) between
                      the given rayline and the model couch surface

    AOI   Incidence angle of the given rayline and the local couch surface
                      approximated by one of 43 couch model definition planes
Prerequisites

- Array of the couch model definition points: XYZ coordinates relative to the selected
  couch origin point. See the model description in the previous section ('Treatment
  Couch Model Example')
- The couch VRT_0, LNG_0 and LAT_0 coordinates for the couch origin at the linac's
  isocenter

Algorithm Steps

- Convert the couch model definition points in XYZ coordinates relative to the given treatment plan origin by using Equation 10.2
- Calculate the radiation source S coordinates XYZ relative to the given treatment plan origin using

$$
S = \begin{pmatrix} x_S \\ y_S \\ z_S \end{pmatrix} = \begin{pmatrix} x_{ISO} \\ y_{ISO} \\ z_{ISO} \end{pmatrix} + 100 \times \begin{pmatrix} \sin(GNT)\cos(RTN) \\ \sin(GNT)\sin(RTN) \\ \cos(GNT) \end{pmatrix} \tag{10.3}
$$

The $(ISO - S)$ directional vector of the source S from the isocenter ISO is determined by using spherical transformation and gantry and couch angles so that the source S coordinates can be determined using parametric equation of a line and a line-multiplication parameter corresponding to $|ISO - S| = SAD = 100$ cm.

- Loop for each of the 43 couch model definition planes ...
  - Calculate the given plane and rayline intersection point. From analytical 3D geometry, the intersection point of a rayline connecting the source S and the in vivo dosimetry reference point N and each (k-th) couch definition plane calculates as

$$
\begin{pmatrix} x_k \\ y_k \\ z_k \end{pmatrix} = \begin{pmatrix} x_S \\ y_S \\ z_S \end{pmatrix} + \frac{a_k x_S + b_k y_S + c_k z_S + d_k}{a_k(x_S - x_N) + b_k(y_S - y_N) + c_k(z_S - z_N)} \begin{pmatrix} x_S - x_N \\ y_S - y_N \\ z_S - z_N \end{pmatrix} \tag{10.4}
$$

  - Check whether the intersection point lies between the given plane (4 or 3) definition points, ie, whether it really intersects the couch model surface. A simple test based on triangle areas calculated from the side lengths using Heron's formula can be implemented:
    - Each defining plane consists of 3 or 4 points. The area within the points can be calculated as the area of a triangle defined by the given 3 points, or, two triangles defined by the given 4 points. Figure 10.3 shows the condition to check whether an intersection point X lies within the area of the triangle defined by the coordinates of 3 points using Heron's formula
    - Record the coordinates of the intersection point on the couch model surface. Note, there will always be 0, 1 or (mostly) 2 intersection points after all 43 planes tested
- Calculate the distance of the intersection point(s) Xi to the source S:

```
norm(S − Xi)
```

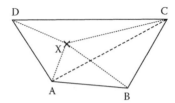

FIGURE 10.3   Testing whether an intersection point X lies within the area defined by the quadrangle ABCD consisting of two triangles ABC and ACD. The area of the triangle ACD equals to the sum of areas of triangles AXD, AXC and DXC. This is not true for the area of the triangle ABC and the sum of areas of triangles AXB, BXC and AXC. Triangle areas are calculated using Heron's formula from known triangle sides.

The closer point to the source S is the couch *entrance* point, the other is the couch *exit* point.

- Distance between the *entrance* point $X1$ and the source S is the *SDD:*

$$SDD = norm(S - X1)$$

- Distance between the two intersection points $X1$, $X2$ is the *couchPathLength:*

$$couchPathLength = norm(X1 - X2)$$

- The *AOI* calculates from the normal vector $n = [a, b, c]$ of the couch definition plane containing the couch *entrance* point (found during the loop above), and the given rayline vector *RayLine = S – N:*

$$AOI = rad2deg(acos(dot(n, RayLine)/norm(n)/norm(RayLine)))$$

In principle, the *couchPathLength* parameter can be used to estimate beam attenuation as discussed previously. It could be also used as a parameter for missing backscatter based on the assumption that the larger the gap, the bigger the effect. However, this problem is not as trivial as missing backscatter, that is, how far a given clinical geometry is from the reference calibration geometry, and is clearly also dependent on other parameters such as field size, beam incidence angle and distribution of patient body on the couch.

The presented MATLAB function script is available in Web Supplement 10.1.

CONSIDERING GENERAL ASPECTS OF IN VIVO DOSIMETRY

Using in vivo dosimetry as an example application of the presented 3D model of the treatment couch top requires some explanation considering state-of-the-art technology and techniques in radiotherapy. Although there are still many treatment plans consisting of a few single fields shaped by multileaf collimators (MLC) possibly with hard or dynamic wedge or even fluence modulation (IMRT) delivered in the world, the contemporary standard is rotational radiotherapy such as VoluMetric Arc Therapy (VMAT), tomotherapy, and so on. Point detector in vivo dosimetry is not possible or extremely impractical for these modern dynamic delivery techniques. However, the concept and general recommendations of in vivo dosimetry are still not obsolete, though alternative and more sophisticated technologies are under development and even available.

Contemporary in vivo dosimetry systems available for modulated rotational techniques are either based on transit detectors mounted on a conventional linac's head, or solutions based on measurements of transit dose through a patient using electronic portal imaging devices (EPIDs). The second principle is also available for tomotherapy solutions. There is an obvious difference in philosophy between the two approaches: the first one monitors the machine performance, that is, the result is independent of the patient, while the second approach is more challenging and the measurement includes both the machine performance and the patient. Although in principle either solution can provide results in a simplified form such as selected points or even just consistency checks of selected parameters measured, the trend is to use the measurement for reconstruction of the three-dimensional dose distribution on a given patient model. Whether this model is the original planning CT or current fraction verification CT (cone beam CT) opens another dimension of the whole problem with respect to method objective and interpretation. However, as even contemporary three-dimensional solutions are used in clinical practice, mostly with the same motivation and even workflow as traditional point detector-based methods, there is still room for establishing really meaningful in vivo dosimetry. A problem is what to do with the data measured and how to interpret the result. Surely, the original motivation of in vivo dosimetry to uncover gross errors in machine performance, plan parameter transfer, and/or patient setup is becoming less and less relevant, especially with automated machine performance logs, radiotherapy verification systems, and mainly pre-treatment plan verification QA and (daily) image-guided radiation therapy (IGRT). Considering the most common practice of performing dosimetry in vivo on patients just for one or a few selected fractions, one can definitely ask the question: what is the actual benefit of such an approach when treatment plan transfer and delivery is verified as part of plan QA process before the first fraction, and patient position and conditions are verified daily using IGRT?

So, in conclusion, unless there are widely available methods for in vivo dosimetry that provide daily reconstructed delivered dose distribution of modern delivery methods with treatment planning system class accuracy so it is possible to monitor cumulative 3D dose and use this for adaptive radiotherapy approaches, the cost-benefit of in vivo dosimetry remains questionable. On other hand, commercial solutions available today already, can be certainly considered as a major step towards the ultimate goal described in the previous paragraph (Mijnheer 2017a,b).

REFERENCES

AAPM TG 105. (2007). Report of the AAPM Task Group No. 105: Issues associated with clinical implementation of Monte Carlo–based photon and electron external beam treatment planning. *Medical Physics* 34, 4818–4853.

AAPM TG 176. (2014). Dosimetric effects caused by couch tops and immobilization devices. *Medical Physics* 41(6), 061501.

ESTRO. 2001. *Booklet No. 5: Practical guidelines for the implementation of in vivo dosimetry with diodes in external radiotherapy with photon beams (entrance dose).* Brussels, Belgium: ESTRO.

IEC 61217: 2011. (2011). Radiotherapy equipment – Coordinates, movement and scales. International Standard by International Electrotechnical Commission. https://webstore.iec.ch.

Jabbari K. (2011). Review of fast Monte Carlo codes for dose calculation in radiation therapy treatment planning. *Journal of Medical Signals and Sensors* 1(1), 73–86.

Mijnheer B. (2017a). Clinical 3D dosimetry in modern radiation therapy. *Imaging in Medical Diagnostics and Therapy Series.* London, UK: CRC Press.

Mijnheer B. (2017b). EPID-based dosimetry and its relation to other 2D and 3D dose measurement techniques in radiation therapy. 9th International Conference on 3D Radiation Dosimetry. *Journal of Physics: IOP Conf. Series* 847.

Vanetti E, Nicolini G, Clivio A, Fogliata A, Cozzi L. (2009). The impact of treatment couch modelling on RapidArc. *Physics in Medicine and Biology* 54(9), N157–N166.

Varian Medical Systems, Inc., 3100 Hansen Way, Palo Alto, CA.

Index

Printed and bound by CPI Group (UK) Ltd, Croydon, CR0 4YY

24/10/2024

01778290-0007